T0200863

Product Lifecycle Management (PLM)

Product Lifecycle Management (PLM)

A Digital Journey Using Industrial Internet of Things (IIoT)

Uthayan Elangovan

CRC Press
Taylor & Francis Group
Boca Raton London New York

CRC Press is an imprint of the
Taylor & Francis Group, an **informa** business

First edition published 2020
by CRC Press
6000 Broken Sound Parkway NW, Suite 300, Boca Raton, FL 33487-2742

and by CRC Press
2 Park Square, Milton Park, Abingdon, Oxon, OX14 4RN

© 2020 Taylor & Francis Group, LLC
First edition published by CRC Press 2020

CRC Press is an imprint of Taylor & Francis Group, LLC

ISBN: 978-0-367-43124-2 (hbk)
ISBN: 978-1-003-00170-6 (ebk)

Reasonable efforts have been made to publish reliable data and information, but the author and publisher cannot assume responsibility for the validity of all materials or the consequences of their use. The authors and publishers have attempted to trace the copyright holders of all material reproduced in this publication and apologize to copyright holders if permission to publish in this form has not been obtained. If any copyright material has not been acknowledged please write and let us know so we may rectify in any future reprint.

Except as permitted under U.S. Copyright Law, no part of this book may be reprinted, reproduced, transmitted, or utilized in any form by any electronic, mechanical, or other means, now known or hereafter invented, including photocopying, microfilming, and recording, or in any information storage or retrieval system, without written permission from the publishers.

For permission to photocopy or use material electronically from this work, access www.copyright.com or contact the Copyright Clearance Center, Inc. (CCC), 222 Rosewood Drive, Danvers, MA 01923, 978-750-8400. For works that are not available on CCC please contact mpkbookspermissions@tandf.co.uk

Trademark notice: Product or corporate names may be trademarks or registered trademarks, and are used only for identification and explanation without intent to infringe.

Library of Congress Cataloging-in-Publication Data

Names: Elangovan, Uthayan, author.
Title: Product lifecycle management (PLM) : a digital journey using industrial internet of things (IIoT) / Uthayan Elangovan.
Description: First edition. | Boca Raton : CRC Press, 2020. | Includes bibliographical references and index.
Identifiers: LCCN 2019057867 (print) | LCCN 2019057868 (ebook) | ISBN 9780367431242 (hardback) | ISBN 9781003001706 (ebook)
Subjects: LCSH: Product life cycle--Data processing. | Internet of things--Industrial applications.
Classification: LCC HF5415.155 .E43 2020 (print) | LCC HF5415.155 (ebook) | DDC 658.5--dc23
LC record available at https://lccn.loc.gov/2019057867
LC ebook record available at https://lccn.loc.gov/2019057868

Visit the Taylor & Francis Web site at
http://www.taylorandfrancis.com

and the CRC Press Web site at
http://www.crcpress.com

"It is not the strongest of the species that survives, nor the most intelligent that survives. It is the one that is most adaptable to change." – Charles Darwin

Dedication

To my parents who educated me about the qualities of self-control over and above honors of education and learning.

Contents

List of Figures

Preface

This book aims to provide a systematic summary of the essential topics of product lifecycle management (PLM) accompanied by Industrial Internet of Things (IIoT) in the era of Industry 4.0. This book primarily focuses on the new product development and new product introduction professionals, who want to start the digital transformation journey from smart PLM through smart Manufacturing in the manufacturing organization. It has been composed to accomplish the minimum demands of implementing the PLM with IIoT program to pave the path to smart connected product.

My target audience will ultimately discover this publication as a useful resource for acquiring expertise in smart product lifecycle management with smart manufacturing. Manufacturers are transitioning from product-driven revenue to solutions and software program, and for making this strategy to work they require to develop and deliver smart connected products. Engineering needs PLM and computer-aided drafting systems to establish brand-new smart, connected products, and manufacturing requires enterprise resource planning to handle all operations in the production line. With the power of PLM, manufacturers can enhance the efficiency of each stage of the product lifecycle by including strategy, execution, tracking, and also preparation. The secret is uniting product information with operational technology dynamically to obtain the understanding that will certainly optimize the product lifecycle. The introduction of sophisticated Industry 4.0 capacities, such as real-time details collection on top of customer-centric analysis, is taking the smart connected product growth to new altitudes. To completely realize the capacity of IIoT and utilize the product information effectively, manufacturers have to understand the challenges caused by PLM and also take the needed steps to resolve them faster in this smart connected world.

In my career as PLM and IIoT specialist, I have had many discussions with product design groups in discrete manufacturing as well as with OEMs. Sharing of data has actually been an all-natural expansion of sharing ideas, goals, and success for those people who have actually participated in design, operations, and data sharing that generally resulted in smart connected products. As technological innovation opens up new possibilities, manufacturers are required to take into consideration new techniques to yield value and earnings from PLM through IIoT.

Work experience is an exceptionally valuable opportunity; it has provided me the chance to have a look at the global journey in a way that otherwise would not have been possible to access in this era. I have really got several amazing opportunities essential in work experience. My entire professional experience opened my eyes to various other possibilities, among them being this extraordinary publication.

I wish you enjoy your reading.

Acknowledgments

I would like to express my heartfelt gratitude to all those individuals who saw me via this book and to all those who offered me assistance and talked things over. Thanks to Cindy Renee Carelli – executive editor, my publisher CRC Press/Taylor & Francis Group – without whom this book would have certainly never ever discovered its method to the digital world, over and above a lot of individuals throughout this global village.

I would like to thank Joel Stein for revealing the course to authoring this book.

I would like to express my love as well as gratitude to my parents, my wife, my son, my good friends, and associates in the business and above all to my well-wishers, without whom this book would not have actually come into existence.

Author

Uthayan Elangovan has 16 years of dynamic experience, which ranges from product lifecycle management (PLM) to Industrial Internet of Things (IIoT) consulting for assorted scope of associations, including automotive, electrical, medical, industrial, and electronics enterprises. He helps and leads PLM, IIoT usage, and subventures and with cutting-edge collaboration tools and techniques, he gives consultations to worldwide clients. Energetic about PLM, IIoT, and its effect on product development guaranteeing PLM, IIoT system meets client deliverables while supporting business process. His interest in making technological advancement in automation influenced him to write this first book *Smart Automation to Smart Manufacturing – Industrial Internet of Things*, which was named as one of the best Manufacturing Automation e-books of all time by BookAuthority. He holds bachelor's degree in mechanical engineering from Kongu Engineering College and master's degree in computer-integrated manufacturing from PSG College of Technology. He currently resides in Tamil Nadu, India, and is practicing as consultant PLM and IIoT, providing business and education consulting through his consulting firm – Neel SMARTEC Consulting. He gives guest lectures and conducts seminars, workshops, and training. He also provides private consultations and may be contacted at uelan@neelsmartec.com.

Visit him at: www.neelsmartec.com

List of Abbreviations

AI	artificial intelligence
ALM	application lifecycle management
AM	additive manufacturing
AML	approved manufacturer part list
AR	augmented reality
AVL	approved vendors part list
BOM	bill of material
BOP	bill of process
CAD	computer-aided design/drafting
CAE	computer-aided engineering
CAM	computer-aided manufacturing
CAPA	corrective action preventive action
CCB	change control board
CE	European conformity
CFT	cross-functional team
CIM	computer-integrated manufacturing
CM	change management
CNC	computer numerical control
DA	data analytics
DCS	distributed control system
DFA	design for assembly
DFE	design for environment
DFM	design for manufacturing
DFT	design for testing
DHF	design history file
DMAIC	define, measure, analyze, improve, and control
DMR	design master record
DTC	design-to-cost
eBOM	engineering bill of material
ECAD	electronic computer-aided design
ECO	engineering change order
ERP	enterprise resource planning
FAST	function analysis system technique
FCC	Federal Communications Commission
FMEA	failure mode and effects analysis
GD&T	geometric dimension and tolerance
HMI	human–machine interface
IIoT	Industrial Internet of Things
IoT	Internet of Things
IR	infrared
IT	information technology
KPI	key performance indicator

M2M	machine-to-machine
mBOM	manufacturing bill of material
MCAD	mechanical computer-aided design
MCD	material content declaration
MES	manufacturing execution system
MPM	manufacturing process management
MRD	market requirement document
MRO	maintenance, repair, and overhaul
MRP	material requirements planning
MRP II	manufacturing resource planning II
NPD	new product development
NPI	new product introduction
NPV	net present value
OE	operational effectiveness
OEM	original equipment manufacturer
OLE	object linking and embedding
OPC	object linking and embedding for process control
OT	operational technology
PCB	printed circuit board
PCBA	printed circuit board assembly
PDM	product data management
PLC	programmable logic controller
PLM	product lifecycle management
PMEA	process failure mode and effects analysis
PPAP	production part approval process
PRD	product requirement document
QFD	quality function deployment
QMS	quality management system
RF	radio frequency
RFID	radio-frequency identification
ROI	return of investment
ROV	return of value
RPA	robotic process automation
SCADA	supervisory control and data acquisition
SCM	supply chain management
SLM	service lifecycle management
SM	supplier management
SOP	standard operating procedure
SPC	statistical process control
SQC	statistical quality control
STL	stereolithography
VR	virtual reality

1 PLM Components

Product lifecycle management (PLM) is a familiar worldwide expression amongst the different manufacturing markets for the process by which a product moves from design through manufacture into the market. The key objective of PLM is to bring together the details, process other than persons attached with the lifecycle of a product within the enterprise. PLM cares for design data in enhancement to processes, constructs and also similarly regulates the cost of product records and additionally store products in a digital documents data source. PLM helps in recognizing product compliance, automates business process through process streamlining and monitoring. It controls multi-user product collaboration over and above availability, throughout the substantial enterprise. Understanding the different capabilities of PLM modules is a vital activity. Enterprises should be proficient in advanced manufacturing, which requires to integrate PLM system and acquire integrated data throughout the supply chain to achieve success. CAD data along with PLM quickly shares and additionally visualizes product details with essential decision makers in a selection of divisions, from engineering to excellent quality control.

Execution of a PLM system will unquestionably be unique to each enterprise, there are core PLM fundamentals that will regularly remain similar. PLM can be deemed both as an information strategy and as an enterprise approach. As an information strategy, it builds a meaningful data framework by combining systems. As a venture strategy, it allows global enterprises function as a solitary team to design, build, manufacture, support and obsolescence for products, while recording best lessons and methods discovered in the process. It encourages the enterprise to make combined, information driven choices at every phase in the product lifecycle. While information includes all mediums – digital and other formats, PLM is largely about managing the digital representation of product information. Based upon individual experience for many years, PLM options can sustain a broad series of products. Examples consist of manufactured parts/products, not limited to such as vehicles, computer systems, electrical home appliances, smart phones, and so on. Numerous products today additionally contain software, firmware, on top of digital components whose information should be taken care of. Some enterprises have long-lived possessions that require to be handled such as utility distribution networks, for example, power, telecommunications, oil and gas, and so on. Enterprises across several commercial markets have efficiently used PLM solutions to handle part/product information throughout the lifecycle for every one of the products manufactured.

The PLM view have improved from handling mainly the mechanical aspects of a product interpretation to consist of the electronics together with software application aspects that have become a better section of numerous products. That expansion remained to push the understanding of what blueprint is incorporated. PLM includes the management of all product-related information from demands, design, deployment, and release to manufacturing. This information ranges from presenting demands, product specifications, and inspection guidelines, to maintain setup information. PLM facilitate products information from various authoring tools and other systems to develop product planning. At the same time, the lifecycle began to include production-focused features with product data information.

Today, PLM encompasses significant areas of process that includes not only programs but also projects with product data management. It comprises of the procedures required to design, develop the product and manufacture in the shop floor and decommission it at the end of its life. PLM utility specify, perform, gauge, and take care of the vital product-related business process. Product data along with functional process strategies are considered a fundamental component of PLM.

PLM FACILITY IN AN ENTERPRISE

To create new products, enterprises need a way to organize each of the associated data. From design sketches, concepts, notes to manufacturing process instructions for manufacturers, enterprises require a main repository for the dynamic information. PLM has actually ended up being essential as the single resource of reality for managing all elements of the product from preliminary development right through product retirement. PLM is the business process automation technique which is used to gather information, people, and process in touch with the lifecycle development of a product. Product development processes and productivity are completely connected and also have a direct impact on the manufacturer's capacity to generate continual company growth. PLM integrates individuals, information, procedures in addition to legacy systems and provides product structure for enterprise and their extended enterprise. At an ever-increasing price, manufacturers have actually pertained to PLM as the prospective solution. Doing this accordingly involves numerous benefits such as lesser manufacturing errors, fewer cycle versions, and inevitably enhanced price to market. As PLM emphasize itself with the entire lifecycle of a product, from conception to customer, it is preliminarily important to recognize the concept of the product lifecycle along with the process. PLM describes the enterprise Bill of Material (BOM) and also furthermore implied alterations concerning the BOM, in enhancement to maintain the background of all previous product development ranges. PLM information is typically developed to take care of engineering along with all cross-functional team (CFT) members of the new product development (NPD) as well new product introduction (NPI) group.

> PLM – A business strategy approach helps to share product information across the NPD/NPI team members, and leverage business knowledge for the development of products from dawn to dusk across the extended business.

PLM develops an incorporated digital platform to:

- Maximize collaboration along the lifecycle throughout enterprises.
- A single source of product management system to sustain varied information needs, to guarantee that the suitable individuals see the ideal product information at the correct time, in the correct context.

- Make the most of the lifetime worth of manufacturers, business product profile.
- Drive top-line income through repeatable procedures.

Businesses that are seeking to get a competitive advantage will do so by boosting their products and streamlining their processes. PLM solutions have actually confirmed to be expensive and too comprehensive besides challenging to take account of existing practices. Why do these solutions, which should be making things much less complicated, generally fail? Have a perfect result as PLM is most likely to aid product design and obtain their computer-aided design (CAD) files regulated, process engineering change orders (ECOs) quicker, increase component reuse, and so on but what will the overall company effect be? By carrying out a PLM system, business can simplify and also reduce each stage of the product development process, which is considered as a crucial success variable. Recognizing the difference in addition to greater possibility for PLM that how it is most likely to add worth to enterprise. An efficient PLM program, with well-defined connected campaigns and a purposeful carrying out plan, sets the ground for an effective transformation.

Establish authorization of PLM based on the roles as per NPD team, setup certain level of access policies for usage of the application and administration, from individuals to administrator who will be utilizing PLM in the office. The efforts demonstrated in PLM are to facilitate training, provide proper access to customer collaboration in the NPD/NPI process as very early as possible. Acknowledging that there could be obstructions along the road leading to PLM and correspondingly all kinds of stakeholder need to manage including leaders, enrollers, agents, champs, blockers, and accepters. Success is achieved by engaging and taking over barriers over these members, or finding out simply exactly how to make it despite them. Decision makers within the enterprise frequently look towards innovation to overrule these challenges.

PLM data will transfer straight to various other legacy systems and alert the product managers, creating even more specific details in addition to forecasts to the production planning control members. Integrating various manufacturing systems also practically eliminates the opportunity of inconsistencies existing in between them. With a lot more details concerning your process at the fingertips, it helps a lot additionally to easily determine feasible places for improvement. By encouraging information-sharing and communication between departments, system grouping can influence collaboration between departments and thus, will result in unexpected solution rework.

VALIDATE THE COST-EFFECTIVE INVESTMENT

PLM was constricted to designers for protecting design information. As products end up being much extra difficult and additionally technically advanced in the period of the Internet of Things (IoT), PLM must connect to facility, connected products which consist of mechanical electrical components, software program application, digital devices into BOM that attach with various other products, services, social collaboration of things exchange of information by means of mobile innovation within and extended business process in PLM to utilize single source of truth.

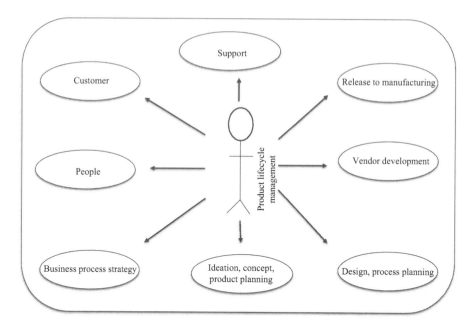

FIGURE 1.1   Product lifecycle management (PLM).

In addition to the needs pointed out, I now add the following few factors as well.

- Prompt PLM training to business stakeholders will certainly make PLM establishing dynamic and also it will connect the space between the individuals and the PLM professional.
- Quarterly business review to obtain a much deeper understanding of the solution in addition to future approaches to plan PLM roadmap.
- ROI calls have to be established.

COMPONENTS OF PLM

PLM starts at the concept phase where quality, trustworthy parts with foreseeable, lengthy lives are chosen. Once the specification and validation of the product begins, all vital components are securely regulated through an enterprise business process. PLM includes all the critical areas of product design and development of data, along with supplier, change management, and so on. This makes it possible for the enterprise to address the core area of robust product development.

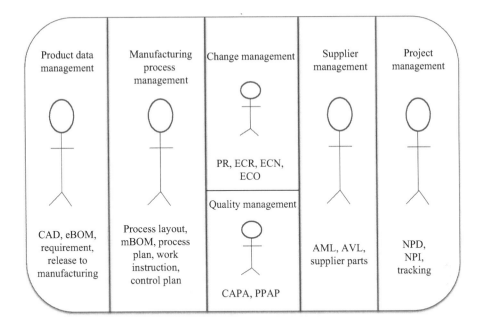

FIGURE 1.2 PLM components.

PRODUCT DATA MANAGEMENT

> Product Data Management (PDM) – A system utilized throughout product design to shop and also obtain information to make sure that the details are intact throughout the life cycle of an item. When they need it, it benefits the usage of simultaneous engineering while maintaining control of information and also dispersing it immediately to the individuals who need it.

The PDM system is utilized during product growth to store and also obtain information to see product details and consistency throughout the life cycle process. When required by the cross-functional team, it benefits by using simultaneous design while keeping control of information and spreading it quickly to the people who need it. PDM was particularly developed to be extremely easy to execute, easy to utilize, budget-friendly for midsize to small manufacturers. PDM aids enterprise in a method to manage CAD data. Traditionally, this was a solitary intranet server taken care of within the enterprise. People can access records from throughout the geography precise like that of PLM. PDMs benefit variation control and normally utilize a check-in/check-out method. Every document gets "had a look at" when it is being modified. When it is being changed no person else can modify that document. When it is done being modified, it obtains "taken a look at" back into the vault for somebody else to change. Every change is tracked, due to the fact that the variant control is streamlined, it is straightforward to return to a previous version. PDM is the core of PLM.

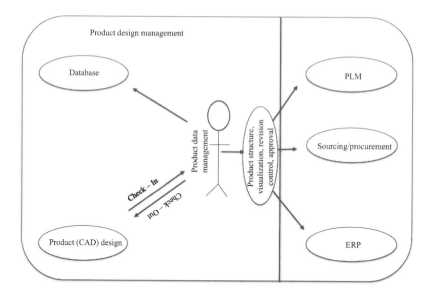

FIGURE 1.3   Product data management (PDM).

MANUFACTURING PROCESS MANAGEMENT

Manufacturing process management (MPM) defines a concept of applying enterprise business process management tools to the locations of factory and supply chain activity administration within and across the comprehensive venture. Product structure manages eBOM, whereas MPM integrates eBOM with bill of process (BOP) or mBOM.

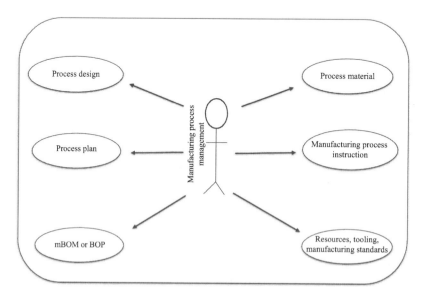

FIGURE 1.4   Manufacturing process management (MPM).

MPM systems are a process-focused solution with a solid connection to product along with production floor/resource information. The basic goal of MPM systems is their ability to import the BOM with alternatives and model-mix content, in addition to map the production procedure information to the components on the eBOM which are developed. MPM data model is process driven with bidirectional and solid associations to the product's eBOM. Along with the product and also process data information, plant source data stores that create the basis of the MPM. Manufacturing/Process planners compare making alternatives to improve the production process, examining out variables such as particular production lines and also simplify activity within a factory; thereby the production strategy is recorded in a manufacturing process planning system using CAD/CAM software application, along with work instructions or manufacturing process instructions are prepared for the operator. In any kind of PLM system, the MPM process serves as an important bridge from product format to manufacturing execution. The electronic manufacturing part of the PLM procedure acts as a bridge from product format to making preparation work, and additionally on source in addition to stock scheduling. Whereas the emphasis of countless design driven processes jumps on specifying "what" the part/product is, MPM focuses on defining "precisely just how" to produce it. Manufacturing process management is the last frontier in the pursuit to electronically connect product design to production in an initiative to improve information quality and reduce time to market.

CHANGE MANAGEMENT

Changes are unavoidable throughout the product development process, as it is essential that these modifications be taken care of effectively with a cross-functional group of professionals within the company to stay clear of squandering time, money and also resources on inaccurate parts. As product intricacy rises as well enterprise end up being more dependent on their supply chains, handling the changes and also interacting change status across several teams inside the enterprise and right into the supply chain is a raising challenge. Change management (CM) is an essential role in the product manufacturing industry that aids to keep order throughout the flurry of adjustments that happen in a life cycle of a product. Efficient CM involves product modifications redlining, version changes, effect analysis, adjustment orders, and workflow authorization procedures.

CM consists of problem report, change request, change notice, and change order. In short, engineering change orders (ECO) are utilized for adjustments in assemblies, documents, components and any other components related to product design and manufacturing. Change Control Board (CCB) is a board that makes decisions regarding whether proposed modifications to a change task must be implemented. Affected items are components, records, procedure plans, or various other revision-controlled things that will be affected by the modification. These are contributed to a job in the implementation plan of a change notice. When an individual completes the designated job, generally these components are affected. Change could be anything from an adjustment in materials, design adjustment, manufacturing procedures that transform the type, fit or function of the end product. Every modification should be appropriately evaluated by all the cross

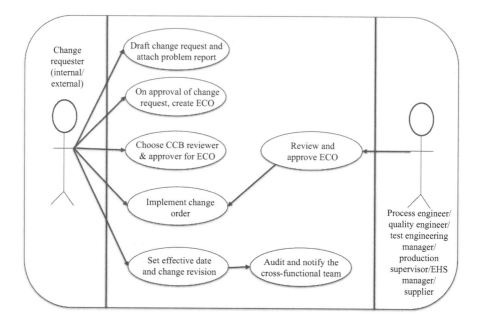

FIGURE 1.5 Engineering change order.

functional department along with quality assurance team to figure out the effect of the modification and carry out the modification through informing the customers. Change management within PLM gives a disciplined procedure to execute a change effectively with the least quantity of disruption.

One of the classic examples of implementing change process is while submitting part warranty to the customers via ECO as international quality standards requirement for manufacturing that keep on enhancing with most of the industrial product manufacturers calling for their suppliers to utilize production part approval process (PPAP) while implementing change to the product or process. The PPAP process should satisfy customer's requirement of producing a new part or product, ECO needs to be implemented along with the documents authorizing the modification and requesting needs to be included in the PPAP package. A customer may request a PPAP at any time throughout the life of a product. For suppliers, this means preserving a quality system that develops, records every one of the requirements of a PPAP submission. It certifies that all areas of the design and production process have been assessed extensively to guarantee that high-quality product will certainly be enabled for customer delivery. Part submission warrant is a recap of the whole PPAP submission that specifies the declaration of material, associated drawing changes, part number, inspection results, and any deviations that are covered by the PPAP submission.

PLM automates and enhances ECO procedures, allowing work to take place concurrently. The NPD/NPI team members can concurrently assess changes, perform analysis, and mark-up changes digitally. As well, it helps to track the history of the open

ECO, where it is pending, and to record the current status. Change notifications intimate the NPD/NPI staff members that there is a new illustration revision and it advises cross-functional team members regarding the change that has to be implemented.

QUALITY MANAGEMENT

Manufacturing process in addition to product high quality are connected to each other; they rely on reproducibility to achieve the preferred goals. Enterprise simplify the product growth process by implementing a PLM system. For any manufacturer, maintaining the quality of a product is the primary emphasis. To make certain continued success, the system needs to be supported by quality procedures created to identify, correct and maintain standards. Quality management (QM) systems allows compliance with the most stringent top-quality standards, including the automotive sector (TS), medical devices (FDA, 21 CFR Part 11), ISO requirements, and correspondingly reduced cost of quality. QM is a reliable system for managing high quality equates an enterprise's purpose on top of goals right into plans and also resources which aid every participant of the enterprise to embrace standard operating procedures.

QM system helps an enterprise to move on in the direction of constant improvement initiatives and also take on information for proof-based decision-making. Quality assurance team members involved in the product growth procedure, tracking with applying quality information throughout multiple manufacturing plants is a formidable task. Design data are no more manageable to continue to be idle in separated silos of business applications. PLM integration with QMS increases the efficiency of mitigating high-quality monitoring issues as early as feasible in the design procedure. PLM supports high-quality procedures to incorporate people, data, and process; these capabilities make PLM appropriate for manufacturing enterprise to improve top-quality product. Enterprises need standardization to generate consistent results, including adequate flexibility for constant improvement to develop a quality-driven culture within the PLM system.

Enterprises deal with decrease in profit margins along with concern in NPD/NIPI team by adopting PLM system and use cutting edge manufacturing technology to manufacture a product, but continue taking care of quality procedures manually using fundamental tools like spread sheets causes costly quality-related risks that frequently evokes product rejections. Integrating quality management systems directly with the PLM processes enables companies to record and also maintain quality documents in a way that reveals adherence to regulatory requirements. The integration process begins with the facility of a cross-functional team responsible for benchmarking present efficiency in connection to the optimal circumstance. PLM to QM assimilation increases the paperwork of essential corrective action and preventive action (CAPA), design history file (DHF), design master record (DMR), failure mode and effects analysis (FMEA) and process failure mode and effects analysis (PMEA). An end-to-end sight of top-quality administration increases development while keeping prices reduced and enhancing performance. From a cross-functional team viewpoint, complementing PLM systems with QMS provides an enterprise the benefit of closed-loop FMEA, CAPA, and other processes. The enterprise's action to quality process documents in the PLM system will certainly make specific compliance as per the standards.

SUPPLIER MANAGEMENT

Suppliers are enterprises that market products and additionally provide solutions to another company by maintaining its own products. Supplier management (SM) is the process that makes sure that value is obtained for the cash that a business spends with its providers. In other words, SM allows manufacturers to take care of and incorporate supply chain data within PLM, extends its capability in the part selection process by making the manufacturer and vendor information offered early in the design phase. Supplier management makes it possible to firmly share product details, including CAD variations and additionally demands without running the danger of direct exposure of copyright. A solid process can help an enterprise in choosing a possible supplier. The cross-functional groups help the sourcing team in confirming whether the selected provider's part/product fulfills the design demands, their manufacturability, production of the BOM. The procurement/sourcing group's skill play an important role in choosing the suppliers from the checklist of prospective vendors.

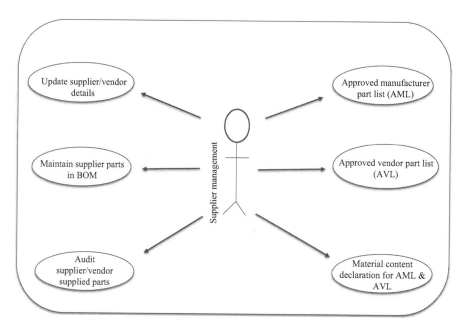

FIGURE 1.6   Supplier management (SM).

Manufacturing enterprise do preserve data sources such as AML (Approved Manufacturer Part List) or AVL (Approved Vendors part List) or sourcing database. AVLs and AMLs consist of the list of alternative vendors and the equivalent component numbers for each element of the main product. The product manager/

designer with a vendor development team make use of the SM module in a PLM system to seek OEM parts that have actually accepted the manufacturer or vendor components in addition to its sourcing status of the supplier components in the BOM. Sourcing/Component Engineer will ask for a material content declaration (MCD) from a vendor, and afterwards attaches the MCD record to the supplier parts on the AML or AVL listing. The MCD ensures that there are no banned materials in their part or product and are certified with environmental regulations such as, EU RoHS, REACH, and WEEE.

SM help businesses to track vendor elements, it enhances the part choice procedure by making it possible for the manufacturers to collaborate with suppliers as a growth of internal resources, making it possible to eliminate design iterations and increase top quality. The supplier information is gathered, kept and updated on a timely basis which is required, right, current and also reliable to help with the total performance of the purchase process. Including a completely digital supplier management procedures to your PLM system will certainly increase the enterprise's design-to-source process. By embracing a sturdy PLM system with a significant supplier management, the enterprise can witness incredible advancement. Reliable partnership, interaction with the external vendors is extremely essential along with audits on a routine basis with the vendors for effective supplier partnership in the NPD/NPI process.

PROJECT MANAGEMENT

The worth recommendation of the PLM system can better be improved via the unification of project monitoring methods. It can be leveraged to handle new product development activity due to the fact that the development of CAD and also the approvals related to this documentation are systematized within the PLM system. For example, a project manager can track the status of the approvals connected to the NPD and NPI as well track the status of the different phases of the product growth with the assigned stakeholder entailed in their project. By leveraging the PLM system to control additional data such as control sheet, process sheet, job instruction, examination reports and also PPAP reports, the project manager can gain insight right into concerns that may prevent their team from achieving the goals they require to accomplish. With coverage, these understandings can be brought onward earlier in the job timeline enabling the job manager to mitigate challenges before they adversely impact the project.

LEVELS OF PLM

PLM's significant function is to remove and establish waste in the product production process. The benefit in taking advantage of PLM in modern-day technology is that the manufacturer can improve the entire management of creating, developing, building, supporting, and servicing a product. Take into account the very first activity to share information and handle product designs, data, and BOM details with PLM. Use basic process besides modifying business procedures to simplify effectiveness throughout the enterprise.

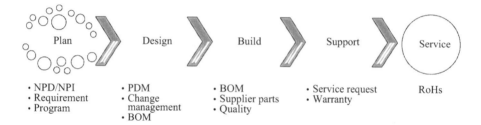

FIGURE 1.7   Levels of PLM.

The PLM system operates with CAD devices as well and integrate to different other business systems like ERP, MES, and so on. It is important to recognize what the enterprise wants from a PLM as a solution, one of the most demanding factors is quality and service proficiencies. The needs will usually break down into the following areas as implementation, adaption, execution, and assistance. Maintaining product data throughout business boundaries as versatile, certified with all stages of the product lifecycle process have actually made a PLM an important roadmap of manufacturing industry.

FEASIBLE VALUE OF PLM

The worth of PLM is the sum of all favorable and unfavorable modifications to the revenue specifications split by the PLM financial investment. Incomes can be gained by either elevated earnings or lowered cost. Revenues in turn may be elevated by increasing the price or quantity, besides expense could be improved as an outcome of improved features or elevated quality. Cost, one of the numerous other hands, may be reduced by decreased expenditure of product or individuals. Ultimately, the price of people can be lowered by decreased time or lowered cost. The PLM might have a positive impact on many of these variables. A great deal of much more reliable searches for digital product information may reduce the moment for product advancement. Enhanced top quality of digital product details can cause enhanced online reality simulations along with for much more trustworthy as well as far more trustworthy information sharing.

A PLM solution is a coherent combination of business objectives, techniques, process, CAD systems, computer-aided manufacturing (CAM) systems in addition to computer-aided engineering (CAE), and PDM systems. The term PLM application consists of the tasks of developing and deploying a PLM service. The PLM applications are commonly done with one or several jobs. The outcome from these projects is the PLM service, which is tentative prior to application, and actual succeeding to implementation. PLM is a broad idea without an accurate meaning of its borders, concentrating product information details. The concept is comprehended with not only CAD, PDM systems, yet additionally a series of other systems as well. Those elements consist of business process, stakeholders, and development. The real value of PLM advancement does not originate from the personal capacities, nevertheless from their adaptation. PLM helps firms take a cohesive, alternative view of the product-related procedures. It also keeps the product information integrated with supplier firm's beneficial efficiencies and also assists in far better deal with the effect of adjustments. It assists enterprise care for the complexity of today's product, consisting of the change in the direction of smart, systems-oriented things that incorporate mechanical, electronic, and as well as software application capabilities.

Provided the different modules of PLM, with the advancing and broadening nature of the innovation there is a whole lot to take into consideration when you are searching for a PLM system, considering that the level, objectives and advantages of PLM executions differ based upon company strategy. Recognizing what PLM is, what it does and also what it can do for business is a very first step in choosing the best PLM software program application for your service.

SUMMARY

Collectively, the purpose of PLM is to interact the information, process as well as people attached with the lifecycle of a product within the enterprise. The PLM assists in recognizing product compliance and additionally automates enterprise business process. PLM is the sum of all favorable and undesirable alterations to the revenue requirements divided by the PLM monetary investment. Profits in turn may rise by enhancing the price or quantity, over and above the cost would be enhanced as an outcome of enhanced features or elevated top quality. If manufacturers have wondered whether traditional PLM can transform the way in the era of smart connected environment with the technological development of Industry 4.0. To get superb high-quality smart product along with smart service to the marketplace promptly, manufacturers need to enhance their product lifecycle process with the digital innovation in this smart connected world. The assimilation of collaborative PLM with various other application systems in a business ecosystem will focus on the growth to enhance the performance of such vibrant collaboration among enterprise to compete in this global village.

BIBLIOGRAPHY

1. PLM | Product Brief | PTC, https://www.ptc.com/en/products/plm
2. PLM | Product Brief | Siemens PLM, https://www.plm.automation.siemens.com/global/en/products/plm-components/
3. PLM | Product Brief | https://www.3ds.com/products-services/enovia/products/
4. PLM | Product Brief | https://www.autodesk.com/content/product-lifecycle-management
5. Aras | Product Brief | Product Lifecycle Management, https://www.aras.com/en/capabilities/product-lifecycle-management

2 PLM Ecosystem

Digitization of the market requires smart product and services. Smart products comprise mechanical, mechatronics, electronic devices elements with an increasing amount of software application program. To get excellent quality products to the marketplace promptly, suppliers need to improve their product lifecycle processes. One considerable obstacle encountered by suppliers is the harmonization of product lifecycle management (PLM). Modern enterprises are dealing with ever-enhancing obstacles of shorter product lifecycles, enhanced outsourcing, mass modification needs, a lot more intricate product, geographically distributed product design teams, supplies subject to fast devaluation, and fast gratification demands. To successfully tackle these challenges in modern-day collaborative enterprise setting, innovative industrial capacities are required in order to obtain affordable advantages in today's connected world.

GOAL OF INTEGRATING PLM WITH ERP

PLM and enterprise resource planning (ERP) look after different company needs; the procedures they endure are carefully aligned with the existing business need, so it is handy to incorporate enterprise PLM and ERP for the product design, development, and manufacturing. The two systems can connect with each other, each making the process more accurate and advantageous. The PLM is more focused on the advancement of the product, and the ERP focuses on the sources needed for production. The continuous increase in customer demand for smart product, makes manufacturers to utilize both systems in the order in which the product have been designed, built, and delivered. By the time when the product is set for production, the PLM will certainly have all the details pertinent to the fact. The details with the PLM may be modified if there is an alteration in the resource preparation work procedure. Incorporating these two systems requires that both of them have the very same information in addition to the most up-to-date details in all times. If the enterprise uses an ERP and not a PLM, there will certainly be a space in the business process automation documents that can make surveillance of the process additionally expensive and difficult.

Product data can transfer straight to various other legacy systems and also alert its analyses, creating even more specific details with forecasts. Integrating both systems also practically eliminates the opportunity of inconsistencies existing in between them. With a lot of more details concerning the business process at the fingertips, it helps determine easily the feasible locations for improvement. By encouraging information sharing on top of communication in between cross-functional departments, system combination can influence collaboration between departments that will result in unexpected solution rework. Details on the product and resources are required for manufacturing, which are tracked as NPD program. Sharing this kind of information in between both systems will certainly make both information resources add great value to the enterprise.

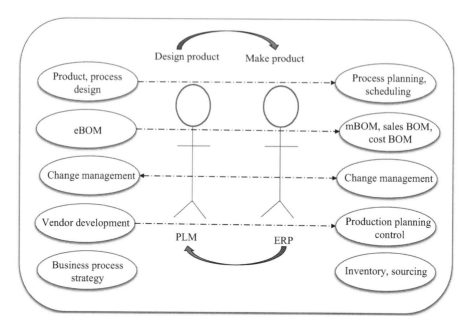

FIGURE 2.1   Integrated PLM and ERP.

For instance, by systematizing technological information in a common expertise base and manufacturing essential data, a PLM allows customers to better comprehend details worldwide and thus foster advancement and also collaboration between all stakeholders, both internally and externally. In addition to the benefits, a PLM provides inherently, it likewise enhances the benefits of an ERP by allowing a unified monitoring of all technological data. ERP, therefore, relies on the understanding of the product along with its environment given by the PLM to optimize the data flow management for manufacturing and logistics in addition have referral product data. ERP without PLM has a hard time to organize product record and might mishandle product changes to be inaccurate with costly rework. PLM is the beginning tool that ERP should follow. PLM is used to manage the design and development of a product, and ERP is used to handle the mass product purchasing as well as resource preparation up for sale along with manufacturing.

INTEGRATION OF ERP AND MES

Sharing data between the production floor and business applications will enable manufacturers to accomplish new levels of effectiveness in lean manufacturing procedures. MES and ERP solutions have been at the center of manufacturing for a number of years. As companies significantly depend on these technologies that continue to be competitive, and as the innovations themselves end up being advanced, manufacturers will reap the advantages of integrating ERP with MES. The MES system is located in between the ERP and industrial control system and has substantial possibility to be efficiently made use of to maximize service processes on the shop floor. Combination of both the systems enables manufacturers to digitize procedures, in which deploying MES is a critical necessity for manufacturers who are leading the journey in the direction of a digital future.

ERP connecting with MES has the prospective to transform manufacturing into a business value. Both systems have actually typically stood alone for many years because they handle different features; however, both are sufficient to be able to develop a nearly ideal system if integrated. Statistical process control (SPC) and statistical quality control (SPQ) components of MES allow manufacturing control to forecast and visually track what needs to be performed in regard to production, in order to keep up manufacturing levels, preserve high quality, and optimize plant performance. To ensure high quality, it is very important to maintain process within appropriate control limits. SPC helps in identifying the main areas of waste and inefficiency in the shop floor, thereby enabling process and product quality to improve faster. ERP outlines standard operating procedures (SOP), step-by-step instructions that are needed to complete a work in the shop floor by the operator that help to maintain safety and efficiency. SOP not only let the operator know how they should be doing their job but also let them know why. SOP in place will also give the option to scale manufacturing process more quickly to MES application. SOP form the basis for ISO and other quality management system accreditation. Finally, it is all about coverage to key data, automating procedures, enhanced quality, increased effectiveness, and enhanced productivity. In the long run, operational data need to flow and make sure shop floor workers' tasks are less demanding, easier, and a lot more efficient.

ASSOCIATION OF ERP WITH MES

MES is the extensive system that controls all the activities occurring on the shop floor. It begins with all the various orders from customers, the material requirements planning (MRP) system, the master schedule, and various other planning sources followed by creating the product in the most reliable, affordable, suitable, and top-quality way possible. ERP will certainly be able to resolve challenges in various aspects, such as resource planning, supply administration, and finance; but modern production atmosphere requires far more than an ERP system alone, even more versatility, data, and connectivity. Bonding of MES with ERP systems enables manufacturers to manage work orders and various other resource needs.

The MES solution is engineered to manage the process that ERP cannot gear up manufacturers with a service that adjusts to their complex transforming production environment. Incorporating real-time data concerning the schedule of resources across the entire supply chain with ERP systems can help manufacturers lessen unnecessary interruptions and further delays. The ideal MES system allows placing machine information, operation information, and traceability data to work to address more important functional difficulties. One of the classic examples of integration of ERP with MRP is the well-organized functioning of engineering change order as any kind of product modifications requested by internal or external customer need be transferred to manufacturing systems as soon as possible to prevent holdups in satisfying customer work orders.

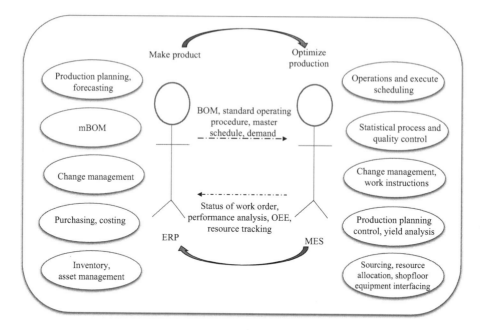

FIGURE 2.2   Integrated ERP and MES.

INTEGRATION OF PLM, ERP, MES

The three standard practical applications to sustain the manufacturing market are ERP, MES, and PLM. ERP focuses on the control of physical standard material, manpower, and machinery/facility needed for producing product, whereas MES connectivity, monitors or handles elaborate manufacturing systems and data streams on the still as PLM integrates people, specification, process, usage, and legacy systems and provides product information structure with enhanced business flow to the enterprise. By securing the technicality in between PLM, MES, and ERP systems, manufacturers wish to facilitate information sharing between the sensible locations of design, development, resources, and shop floor operations. The goal is to deliver coverage of data source that supports improvement of product distribution cycles and remove repeated hand-operated task and also waste, in addition to proactively help provide remedy to establishing high-quality issues before they end up being equally costly and posing barriers to customer fulfillment.

The enterprise data steps throughout the PLM–ERP–MES trinity at the same time, with various degrees of adaptation, at different stages of the product lifecycle, versus numerous details versions in addition to structures. The core concepts in between this combination concern master data management (MDM), product along with business information placement throughout the product lifecycle, and collaboration throughout the comprehensive service. In a nutshell, PLM cares for technical choices, ERP deals with tactical choices, while MES cares for operational options. PLM, ERP, and MES integrate to develop a structure for a contemporary analysis

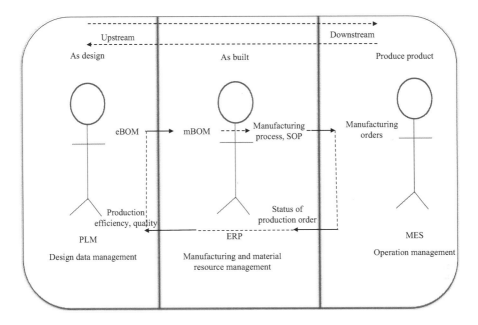

FIGURE 2.3 Integrated PLM, ERP, and MES.

of lean production. The combination of PLM, ERP, and MES makes it possible for manufacturers to share mutual details throughout the venture, which aids in making smarter choices with much better details.

As part of the product-to-production system, that is, PLM through ERP to MES, there are straight links from PLM to the production line, as one of the most reliable instances, which is common across work instructions, transmitting details from the production line to the PLM and vice versa with nonconformance outcomes. One more important capability of MES is that of integrating shop floor equipment; it also has the capability to capture data directly from the manufacturing equipment. The PLM-MES direct integration has the possibility to provide handiness using collaborative method, which have the ability to team up. The level of adaptation in addition to specifically just how lean the business can be constitutes new paradigms to enable competitive benefit.

INTEGRATION OF PLM, ALM, AND SLM

Attending to the present commercial difficulties and proceeding to develop cutting-edge products and solutions, enterprises must prepare for and enhance their sensitivity by taking on procedures that are more nimble, better incorporated, and interoperable with those of their customers or partners. Product manufacturers are facing challenges in providing effective, efficient, and also risk-free service to the customers. Service lifecycle management (SLM) treats manufacturer's service as a full system, which is incorporated right into the product lifecycle as early as possible in order to supply the optimum value to the client and enhance gains for the venture.

Effective service and maintenance activity is connected to product and asset management, which forms a part of SLM environment. SLM enhances product service to the next degree of performance by utilizing information collected in the field to facilitate higher level operations like warranty management, components management, source planning, logistics, as well as pricing. The outcome is enhanced revenue and more profitable solution contracts. Product dependability is the primary driver of service and guarantee expenses. Overall, it improves customer fulfillment as well as lowers repair time and guarantee fixing expenses.

Application lifecycle management (ALM) is a transverse technique based on the procedures of systems engineering in addition to demands engineering, with the objective of assuring conformity to a total system and its specification. ALM is additionally interested in the monitoring of facility systems consisting of a huge quantity of embedded software program. Software source code management capability is integrated very closely within the ALM suite. For smarter, much faster products to be developed, it is vital to merge ALM with PLM.

Examine the real-world product use and condition data throughout products and consumers with PLM, ALM, SLM, and top-quality systems to make it possible for data-driven closed-loop lifecycle administration. In any areas of NPD/NPI, it is essential to be able to identify and also manage the as-maintained setup of fielded products and firmly disperse software updates to enhance efficiency, guarantee compliance, and provide new product variants. Manufacturing enterprises regularly focus on improving outstanding quality, performance, and stability in addition to favorable collaboration with customers. It has really become apparent that complying with the generation of competitors is digital that needs adjustment in everything, from manufacturing product to providing constant service.

MEETING OF PLM AND SLM

The unification of SLM into PLM strategy for a manufacturer ensures that solution is kept in the leading edge of product assistance, shipment, and advancement boosting the earnings of service contracts and making it possible for customers to obtain the most worth for the investment across the lifecycle. Assimilation of SLM into PLM allows the solution enterprise to ensure they are straightened with the authorized components checklists defined by engineering. PLM system feeds data to SLM bidirectional, comprehensive product requirements for field service technicians, and also product issues revealed by those specialists can consequently assist the engineering team to make design improvements. By having the information produced by SLM services conveniently available and not simply considering certain end results, business can discover methods to identify the root causes of the solution problems and establish troubleshooting methods.

Performance evaluation of the product is to understand how components are operating, when they are failing, what configurations they are failing on, exactly how commonly they fall short, and under what conditions they are failing. Without the above remarks, the manufacturers cannot be effective in constant product renovation, successful solution contracts, supplier performance, and long-term customer contentment, thereby undermining the feasibility of the line of product.

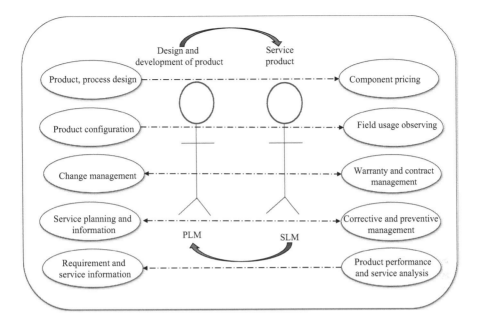

FIGURE 2.4   Integration of PLM with SLM.

Features of SLM

SLM enables taking full advantage of the value of the product by offering worth to the client. Engaging with customers beyond sale opens new path to manufacturing enterprise to deliver value, which will certainly generate income into new profit streams. Additionally, with SLM in position, manufacturers obtain higher understanding right into just how products do with customers, when products require to be serviced, and how accurately to service products. The existing components on top of service operations can be changed from cost to earnings by leveraging SLM to its full advantage of performance. The components of SLM are as follows:

- Components pricing, planning, and forecasting
- Enterprise asset management
- Logistics management
- Service knowledge management
- Warranty management
- Contract management
- Corrective and preventive management
- Service performance analysis
- Returns and repair management

The SLM lets manufacturing enterprise plan their service resources and in minimizing service expense it also lessens avoidable return of any malfunctioning products while boosting operational performance of the solution. The SLM assists in the production of appropriate job guidelines, operation guidebooks, cookbook, precise service, and maintenance procedures for worldwide client market.

MEETING OF PLM WITH ALM

ALM and PLM make use of isolated techniques, operating within the limits of their very own repository. Product technology structure varies from mechanical to mechatronics. ALM and PLM manage complicated regulative conformity and product safety needs in a global smart product development environment. Cutting-edge products along with Industrial Internet of Things (IIoT) are creating a requirement to break down these limitations, fundamentally reshaping the lifecycle management process. Software application is taking control of larger and bigger pieces of core or expanded capability from hardware. Merging of PLM and ALM is essential when developing a system with both software and hardware components.

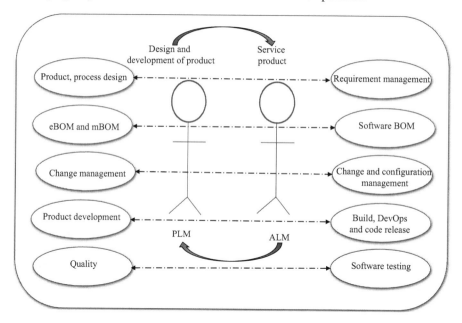

FIGURE 2.5   Integration of PLM with ALM.

Product innovation and manufacturing comply with different advancement streams for hardware and software. These advancement streams progress in total or in partial isolation; this likewise implies that the combination of these two will have to be seamless as well as perfect. In smart manufacturing enterprises, PLM with ALM is considered really important, whereby successful characteristics assimilation of the development of software and hardware systems is required to see if the return on investment is high. To be reliable for manufacturers, ALM operation takes into account the administration, software application modification together with configuration management, system administration, and quality monitoring in a solitary, essential service. The practical basic process follows as integration of requirement and BOM management in PLM, with software requirement and software BOM management in ALM. ALM supplies visibility into product release readiness, supports alternative intricacy, automates growth processes, and ensures full lifecycle traceability.

Features of ALM

Manufacturing enterprises undergo tremendous changes in their business processes, business models, and products constantly resulting in the adaption of ALM as part of the smart product development, wherein ALM is intended to increase the overall quality of the software by handling application development and application operation equally. ALM is the most reliable elaborative strategy to identify the value of the application and also the sources made use of gradually. ALM takes into account governance, development, maintenance, and retiring of software application, that is, concept to end of life. The components of ALM are as follows:

- Requirements management
- Design planning and estimation
- Configuration management
- Change management
- Release management
- Quality assurance
- Deployment
- Maintenance and support
- Audits, metrics, and reporting

By integrating people, processes, and tools from the beginning to finish, ALM enables services to construct better software application as well as manage it easily. Developing an app for smart product without an ALM is like setting out a strategy in hallucination. ALM provides clear instructions for the software product development group, accelerates growth, and helps decision-makers make better choices throughout an application's lifespan. It is an important component of effective enterprises today.

INTEGRATION OF PLM WITH SCM

The quickly changing expectations and consumer assumptions, manufacturing enterprises are regularly striving to minimize product development time. The life cycle of design–release–manufacturing–customer phases in the current production environment is astonishingly complicated due to customer demands. The majority of the information in PLM system relates to design and data management, while a lot of information in supply chain management (SCM) relates to approved manufacturer/vendor, supplied parts used in operations. PLM offers effective supply chain insights because they include comprehensive product-associated data, from design to obsolescence.

SCM involves a consistent flow of exact, timely useful information regarding basic materials, processes, partnerships, and service providers. Supplier details are

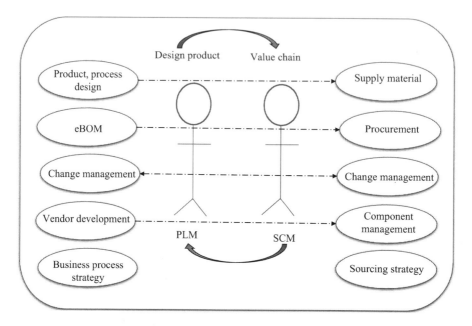

FIGURE 2.6   Integration of PLM with SCM.

taken care of in SCM system where the details are preserved for parts/products that are in prototype phase as well as in manufacturing phase. The NPD/NPI team develops concepts and redesigns new components/parts that are maintained in PLM system as per demands that come from the customer. The product design engineering team has to "make" or "buy" decisions at an extremely early stage in the product lifecycle, which have extensive effect on the supply chain. PLM makes it possible for services to collect precise information from various functions throughout the enterprise supply chain, all the way from design to distribution. Manufacturers require to continuously review internally the supply chains for top quality and also compliance.

WHAT BUSINESS NEEDED TO LEARN ABOUT PLM IN THE PERIOD OF INDUSTRY 4.0?

The presumption is that PLM is a huge enterprise solution in many of the establishments, and cost comes to people's minds promptly, considering that it is good only for the OEMs and discrete suppliers. CAD engineers felt that the problem was that many on the design side saw PLM as a threat rather than a chance.

The global product growth, manufacturing, and end-user with circular patterns are changing quickly such that enterprises require to reflect much more on services that can create the abilities called for to remain to be affordable. Almost every facet of product recognition along with usage is digitized in addition to electronic service; the nature of improvement in usages and likewise products is changing. These adjustments require that product developers and manufacturers tackle brand-new means of thinking with smart connected cooperation with each other in order to define and make products as well as take care of their lifecycle. The capability to team up is essential in addition to visibility and compatibility in order to trade product information effectively. Collaboration is progressively important in light of the above-mentioned evolving modern innovations.

PLM is a proceeding concern, simply like enterprise is progressing. Is PLM possible with modern technology? These are the communications between individuals, business entities, manufacturing sites, and information technology. The emphasis of PLM remains in both product and profile management; it requires administration, enterprise analysis, warranty management, as well as compliance management. PLM implementations are much quicker and less challenging, and the procedures and solution analytics are extra relevant to the target enterprise. Out-of-the-box, template-based techniques give relatively fantastic "common" PLM process moves. Industry 4.0 suggests establishing and producing resilient goods not only for the Internet of Things but also for machines and plants with control devices. Achieving the digital ecosystem is possible by integrating industrial value chain across the end to end lifecycle of the product. The structure of this digitization is no less than PLM.

DIGITAL TRANSFORMATION OF PLM

Technology is a fundamental part of an enterprise. Expert System, Artificial Intelligence (AI), Big Data information, and IIoT are the buzz words across the titans. Man-made technological expertise coupled with predictive analytics can anticipate a trouble and thereafter offer a choice to overhaul it. Establishing a digital thread between the factory-floor and PLM along with on-field services enables real-time interaction between developers and experts. This allows distributors to much better comprehend layout concerns in addition to performance gaps, improving product premium and also performances. The PLM task needs a mindful understanding of enterprise procedures included and called for re-engineering task, with regard to maximizing the PLM technique. It is required to have a global overview of the smartness of business's information processing system. Analytics is critical to optimizing the whole worth cycle, entailing vendor effectiveness, product setting, field remedy monitoring, and client support.

The dynamics of the market are transforming, and therefore is the product complexity. With enterprises manufacturing industries marketing their products internationally, the standards of the competitions are ever changing. Further, each industry has its own demand–supply curves counting in the geopolitical situation at a particular time. Businesses are investing large bucks on equipment monitoring-based solutions to prepare for future. Presumptions are climbing as firms take care of budget-friendly stress to drive preparations and likewise bring products to market faster. Industry 4.0 has in fact triggered change much faster than formerly developed. Manufacturers will definitely require to exit or adjust. PLM is now at the heart of the digital business. The IIoT and AI have considerable possibilities to collaborate with PLM, thus enabling to transact information that build up from each part of the process.

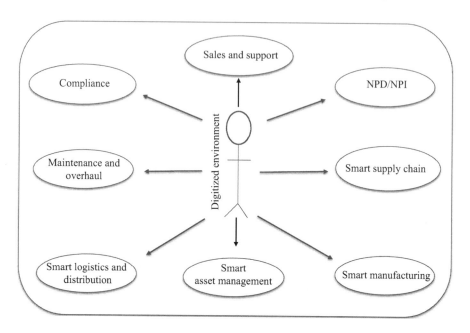

FIGURE 2.7   Digital transformation atmosphere.

SUMMARY

Comprehensive PLM ecosystems will certainly provide a strong structure for traditional product development objectives. The manufacturing enterprise goal is to
achieve digital transformation journey in the smart connected world. PLM has
come a long way and it is expanding considerably, significantly attaching to the
global manufacturing and information communication technology. With the fast-
developing digital enablers, numerous cross-functional team members encounter a
paradigm transformation across the product lifecycle in the near future. The expanding demand for combination by the customers in product development besides the
processes around the manufacturing world will lead to a brand-new kind of interactive experience that opens a lot of new possibilities with Industrial revolution 4.0.
Dynamic enterprises are reconsidering their business models and also possibilities
created by smart connected products. Enterprise needs to establish methods for
how to leverage and also make best use of the IIoT, augmenting PLM with context
understanding as well as even more flexible solutions. Digital improvement journey
will have footprints in numerous layers of the enterprise. Manufacturers are looking
forward toward intelligent manufacturing facilities and enterprise applications that
enable people to set apart from their rivals and also resolve the growing abilities gap
in producing. This leads to the thought that PLM integration with IIoT is going to be
a game changer for manufacturers in the era of Industry 4.0.

BIBLIOGRAPHY

1. Batenburg, R., R. Helms, and J. Versendaal. 2006. "PLM Roadmap: Stepwise PLM Implementation Based on the Concepts of Maturity and Alignment." *International Journal of Product Lifecycle Management* 1 (4): 333.
2. Sääksvuori, A., and A. Immonen. 2008. *Product Lifecycle Management.* Berlin, Germany: Springer.
3. Scholten, B. 2009. *MES Guide for Executives.* Research Triangle Park, NC: International Society of Automation.
4. Park, J.-M., and B.-S. Kim. 2013. "Review of Social Manufacturing Technology on Product Life Cycle Management (PLM) Base." *Journal of Korean Institute of Industrial Engineers* 39 (3): 156–162. doi:10.7232/jkiie.2013.39.3.156.
5. Leon, A. 2014. *Enterprise Resource Planning.* New Delhi, India: McGraw-Hill Education (India).

3 Evolution of IIoT

The collaboration of technological innovations in the fields of information technology (IT), communication, Internet of Things (IoT), and smart connected gadgets is poised to change not just user–machine interaction but also the way in which devices interact with one another. An evolution has its beginnings in capabilities and innovations established by visionary automation providers. One such field where we see this diffusion the most is manufacturing sector. Undoubtedly, energy, healthcare, vehicle, and discrete manufacturers are starting to roll out the Industrial Internet of Things (IIoT), where devices such as sensors, robots, production line machines, and so on are becoming significantly much more connected. The IIoT is the application of IoT in manufacturing industries. It depends on the substantial variety of sensing units that exist in manufacturing industries, along with the vast quantities of data produced by the machines that can be analyzed into workable expertise for much better performance. The possible outcome of IIoT entirely governed by the ability to link automation systems with existing enterprise manufacturing systems and infrastructure. The influence of IIoT is felt most in the manufacturing enterprise across the globe by transforming the methods of various service operations. Enterprises that are accepting the change will see a greater growth graph than the ones who are still cynical regarding Industry 4.0.

IoT IN MANUFACTURING ENVIRONMENT

Industrial automation first started in manufacturing businesses, identifying repetitive tasks by the workers performed in the shop floor that could be handled by machines. It involves the usage of machinery to manufacture an end product with speed, consistency, endurance, and accuracy beyond the capacity of an operator. Industrial machines are driven utilizing electrical, hydraulic, mechanical, pneumatic energy, and computers. Excellent developments have actually been made in automating manufacturing industries. Modern-day automation starts with the introduction of computer-integrated manufacturing (CIM), controlling virtually the entire procedure of developing a product, from concept to design through manufacturing. Over the period, industrial technologists have introduced the use of robots, programmable logic controller (PLC), and supervisory control and data acquisition (SCADA) along with electromagnetic fields to identify and track the tags attached to a product.

Automated lean manufacturing is a proven, powerful method used in production processes, achieved by integrated PLC and SCADA along with the shop floor machines, which helps in boosting efficiencies and reducing waste. Further technological innovation broadens automation from simple, physical processes to data-centric features, consisting of analytics, modeling, and decision-making. Its major emphasis is on making production procedures more efficient and also less costly. Today's manufacturing facilities are home to a good deal of impressive automation, including incorporated manufacturing systems, smart sensing devices, microcontroller systems, collaborative robotics, and so on.

Enterprise has a way of moving toward innovations with the best value. One of the innovations in transformation of manufacturing industry digitized in the era of Industry 4.0 is the use of analytics bundled with IIoT. As the manufacturing industry moves toward digitization, the industrial world is gradually changing service as its data processing grow and provide exceptional insights into how manufacturers can recognize higher financial savings. IIoT intrudes into all aspects of manufacturing and supply chain management for its wide variety of deployment and application to enhance process. Technology has traditionally been used to capture past and present information. Emergence of the IIoT gives lean efforts a major boost in the manufacturing sector. There is a paradigm shift within the manufacturing industries to use analytics to provide better service. At the same time, there is widespread reduction in resources, personnel, and budgets. Analytics does more than creating better situational awareness, leading to cost-savings and increase in efficiency. The IIoT technology brings massive change and a new disruption occurs; enterprise continues with each new technology, thereby enabling manufactured product evolve smartly.

Product data usages are fed back to the new product development or new product introduction teams so that they can analyze the data to make improvements in design and production. The constant exchange of data, integrated with the Industry 4.0 modern technologies that are emerging along with development in information analytics, in manufacturing industries, can lead to truly smart manufacturing facility. The intersection of IIoT with lean technique has the potential to take lean to the next level. The data obtained from connected devices, consisting of customer's field experiences with a range of products, are sent dynamically to manufacturing facilities to enhance production processes as well as minimize waste.

AUGMENTATION OF SMART DEVICE

Nowadays, internet usage has become common; it has touched almost every corner of the world, influencing human life in inconceivable manner. A device or machine detects the environment and data are accumulated with the aid of a sensor positioned. A sensor controls the physical quantity and transforms it into a signal. Sensors transform signal from real life into data and are used in, for example, smart factory, homes, mobile phones, automobiles, and so on.

A sensor is being used as an input to the IIoT application. Sensing units fitted into the best places along the manufacturing floor can instantly provide manufacturing facility supervisors important information and help them take control of process that may go beyond the predefined criteria. The signals can be sent to the desktop in the supervisor's cabin or on his smart device to make sure that he has access to the data from anywhere. The key purpose of a sensor is to pick up the data from the contiguous device and sent them to another connected gadget that is being refined over cloud to take proper action. Different sorts of sensors are being used with IoT tools, depending upon the business need. Automatic sensing units consistently transfer measurements that are additionally a vital component of the IoT.

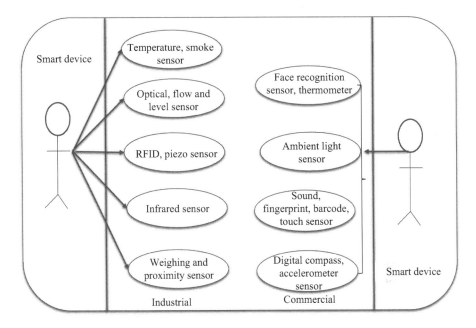

FIGURE 3.1 Technology behind smart devices.

Smart devices have the capability to dynamically adapt to the transforming contexts as well as take actions based on their operating situations. Manufacturers invest resources to research and deploy industrial automation, which is the biggest advantage to the enterprise; thus, IIoT is transforming the way businesses work and compete in the smart connected world. One of the most essential characteristics of the IIoT is context understanding. IoT sensors are primarily small in size, affordable, consume much less power, and know how exactly to harness that data for higher benefits. The IIoT leverages network connections to help with communication between equipment, gadgets, and systems. Manufacturing enterprises have been releasing smart gadgets in numerous brand-new methods to enhance their establishments.

TECHNOLOGICAL DRIVE OF PROCESS

Innovations in industrial technologies are expanding the options to make lower expense and higher worth industrial automation systems. Industrial automation helps in the tracking and monitoring of tools such as pipes, presses, extruders, and assembly lines by the information extracted from M2M and IIoT. Sensible incorporation is made of process and manufacturing features with other business service functions of enterprise in real-time synchronized process, including PLM, ERP, MES, asset management, process optimization, production optimization, supply chain systems, quality systems, and customer care service. The value of closed assimilation of the shop floor operations with enterprise systems is recognized for improving production efficiency. Improvements in the process automation allow the information that physical gadgets create to be evaluated in actual time in order to facilitate self-determination, permit precautionary maintenance, and improve uptime. Machine learning and artificial intelligence along with PLC, SCADA, and distributed control system (DCS) embedded in the business platform create a smart connected business process.

PROCESS CONTROL SYSTEM

Industrial process control includes monitoring and regulating equipment and systems; it refines throughout a multitude of industrial sectors. The role of the process control system is to collect and send information acquired throughout the manufacturing process. It is a simple product with a sensor, often called a main transducer, which receives an input, along with a controller that refines the input, and end in processed output with the help of receiver. The process control system performs efficiently, constantly, and also with as little variation as possible. The two most common control systems to regulate equipment as well as processes, dealing with the numerous analogs, digital inputs, and outputs, are PLC and DCS. The industrial control system is a multidisciplinary system, which takes care of techniques like control systems, communication, instrumentation, electronics, electric systems, and so on.

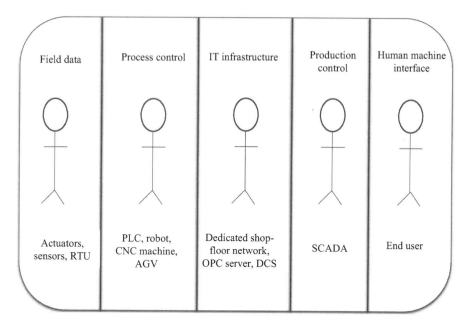

FIGURE 3.2 Process control system.

Industries have begun to move toward integrated plant tracking and control with SCADA systems. It is here that the operator is able to see the whole process from an eagle's eye, provides commands, and observes and responds to alarm systems. The object linking and embedding (OLE) for process control (OPC) is a software interface requirement that permits Windows programs to interact with industrial equipment tools. The OPC server converts the equipment communication protocol made use of by a PLC right into the OPC method. SCADA makes use of both PLC and DCS. All these components handle industrial instruments, smart devices, operational master computer, and dispersed I/O controllers along with HMI suits. It consists of many remote terminal systems for collection of information (obtained from the field), which are connected with master terminal through interaction system, having the primary task of collection of precise data and regulation of process for smooth operation.

OPERATIONAL TECHNOLOGY

Enterprises are seldom successful in accomplishing preferred results from innovation. Operational technology (OT) makes certain that a company efficiently turns inputs to outcomes in a reliable way. The OT is a collection of gadgets developed to work together as an incorporated homogenous system. Industrial systems relied upon proprietary methods and software application; all are managed and checked by shop floor operators and had no link to the outside world. Taking care of computed interactions for industrial applications has actually been confronted with growing complexity of their industrial control systems. To maintain, manufacturers are challenged to develop operation process maturation, specifically a few of the largest adjustments to the OT systems, where they are no more stand-alone systems. If one of these systems falls short, it can have a disastrous domino effect.

OT systems are now connected to company networks to provide operational data and service information. Accomplishing operational effectiveness (OE) depends on the ideal factors of people, process, and modern technology; the option exists within the application of the fundamentals of process monitoring, which has a straight and

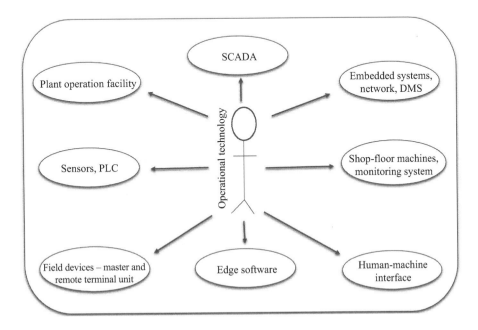

FIGURE 3.3 Operational technology.

substantial impact on operational performance. OE requires that the proper individuals are in place to guarantee that the right processes are utilized, completely sustained by the ideal technology, including tools to carry out the process and support decision-makers. OT is associated with field-based devices that are used to carry out actual operations in the shop floor.

Automation and digitalization of manufacturing process aligning with lean manufacturing has room for improvements and innovation. The basic things that are digitalized in the manufacturing process are operations technology, monitoring, and machine data, which get translated into predictive process resulting in improvement of service maintenance and business intelligence. Prompt accessibility to machine data making use of connected devices to obtain an image of what is happening in real time helps manufacturers in proper planning of operations.

In the smart connected factory environment, control and monitoring of machines, materials, sensors, PLC, SCADA, DCS, CNC, and so on are all possible OTs. In simple terms, OT, commonly related to manufacturing and industrial environments, includes industrial control and supervisory control and data acquisition systems that include collection of devices created to interact as an integrated identical system.

Necessity of Operational Technology

Business may reduce expenses by modifying enterprise process as well removing activities that customers realize it less valuable. The importance of modern technology in business is the capability for computer systems to carry out numerous tasks concurrently. In lots of businesses, survival and also the capability to accomplish critical enterprise objectives are challenging without considerable usage of recent technology. The importance of technology in business is the ability for information systems embedded within manufacturing technology to perform multiple tasks in parallel with less intervention of operators.

INTEGRATION OF OPERATION WITH INFORMATION TECHNOLOGY

The emergence of universal economy, improvement of industrial economy along with business transformation, and the advent of digital environment make IT systems important in today's competitive business. Connecting industrial control system networks with each other and also to the open-air world, it does involve an entire plethora of obstacles from a systems combination point of view, specifically when taking care of networking systems. The network complexity is being driven by a number of aspects familiar to network engineers. One of the main features is the massive activity of venture compute, storage, and legacy, in house and cloud applications.

As industries turn out to be technology-driven, there is a raising demand for skillsets to incorporate essential excellent manufacturing with modern technology. Innovative tools are letting enterprises to develop and examine situations in the digital world, to replicate the design process through production process before a real product is manufactured in real time. Imitating the product-creation stage reduces production time, ensuring the production process supplies what business planned to manufacture. The IT innovation coupled with functional operation technology takes into consideration a significant practical area in business operations. It aids enterprises to obtain a competitive advantage. It provides the foundation for brand-new products, solutions, and leading service that offer enterprises with a strategic advantage. It is a transference, with dispersed intelligence assisting to handle the change from a centralized, unidirectional distribution of data to a platform that enables bidirectional data circulation and automated control. IP-enabled networks connect the OT side at the shop floor to the IT side.

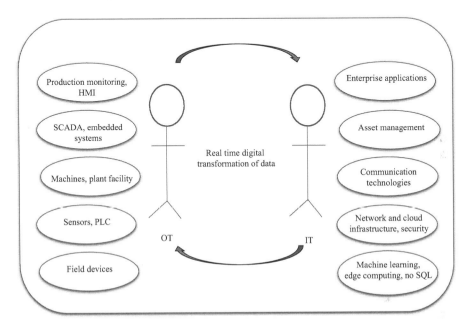

FIGURE 3.4 Integration of operational and information technology.

OT and IT have had relatively separate roles within an enterprise. The product and plant managers in reality operated industrial control systems in a separate network, literally set apart and secured from other areas of their production environment or enterprise systems for safety and security reasons. With the introduction of IIoT, which permits the integration of intricate physical machinery with networked sensing units and software, the industrial machines are made accessible via intranet or Internet to deliver analytics from the data generated to optimize the production process – connection between the OT domain, which entails automation of shop floor, supply chain monitoring, and enterprise asset tracking, where information are produced, and the IT domain, which entails business process automation, where data are consumed. The incorporation of OT with IT is a key factor to achieve smart manufacturing. Emerging technologies like machine learning, AI, and so on, were driving OT/IT integration. OT/IT integration allows straight control and complete tracking, with much easier evaluation of data of the enterprise manufacturing facility distributed across the global village. Integrated OT and IT enterprise results in efficient process and operations.

BUSINESS VALUE STREAM OF IIoT

Enterprises make every effort to establish value for their business and clients. Value is excellent quality, efficiency, and design format regarding product expense. Smart manufacturing is an endeavor to achieve manufacturing intelligence, orchestration, and optimization within production facility. The IIoT develops and accepts new business models, connected processes, and openness among all business stakeholders' participation in the smart production ecosystem. The value stream of IIoT specifies the means of taking care of supply of raw materials and finished products and also takes care of the product data information via NPD/NPI functional departments along with operations through managing the delivery of the end item to the end consumer.

Traditionally, in product manufacturing industries, business strategic value is implied recognizing long-lasting customer demands and producing well-engineered services. In a smart connected manufacturing, products have updates, new features, and functionality that are updated to the customer on a regular basis. The capacity to track product in field operation makes dynamic response to customer; thus, IIoT value will pave path for enterprise to generate income from the manufactured product and customers get its value from the smart connected product.

The IIoT guarantees to supply manufacturing enterprises with huge amounts of information to drive smart intelligence. Yet the value of the innovation exists in exactly how manufacturing personnel utilize it, not in the technology itself. In manufacturing business terms, the worth offered to business are machine productivity, manufacturing facility, uptime output, worker safety, and so on. With regard to manufacturing processes, some of the key service value that generates direct yield outcome are supply chain, operation efficiency, inventory monitoring, and predictive analysis.

END-TO-END IIoT ENVIRONMENT

The IIoT is a network of smart machines that are interlinked to each other to communicate by means of M2M over Internet or intranet and is regulated using a mobile application. A complete IIoT system integrates four unique parts: sensors/devices, connection, and information handling, together with an end user application. The IIoT consists of industrial manufacturers that make use of sensors with communication technologies like Wi-Fi for monitoring, controlling, and remote tracking. The IIoT offers fresh opportunities of customer experience along with user interface that allows prompt decision-making by the manufacturing executives. Sensing systems of the IoT/IIoT system accumulate information from the machines along with the other information from the process control system and send it further refining in the edge software that specifies the different communication methods to develop the data transformation efficiently.

The information developed from IIoT-enabled machines end up being of worth; it just brings details of analytics into the picture. The data analytics (DA) is a procedure that has the advantage of looking at both large and small data collections with differing information to eliminate actionable understandings in addition to purposeful judgments. Smart connected devices are developed in such a manner that they tape-record the data that are used in the day-to-day activities in the manufacturing environment. Enterprises have to make use of the excellent high-quality details available and thus boost their process through cutting-edge innovation systems. The future of IoT is far more remarkable, where billions of things will certainly be interacting with human in addition to various other systems and will wind up being the best. The IoT will bring macro-adjustment in the method we use. In a nutshell, IoT aims to present and resolve all potential challenges online to give a safe and comfortable life for humans.

BUSINESS USE CASE

Realizing substantial gains from connecting newer equipment, oil and gas markets are considering what can be acquired from deploying IIoT modern technologies for collecting and examining data from heritage assets. Connecting the equipment would assist in protecting against device failings, subsequently preventing costly downtime and aiding forecast when the equipment would certainly need maintenance. The service of drawing out and transporting oil and gas is loaded with difficulties. To stay affordable, companies in this sector have to consistently aim to generate petroleum as well as fine-tuned products at a lower price. They are also regularly aiming to improve and prolong the worth of their existing properties while likewise looking for brand-new oil and gas reserves. Also, ecological requirements are becoming increasingly strict, needing openness in operations and also tighter controls on production and distribution.

Obstacles dealt with by oil and gas sector are cost containment stress and connectivity problems with remote oil wells.

- *Required for Ongoing Expense Optimization:* Their target was to take advantage of technology to financially expand the useful lives of oil wells. Provided the manufacturing volumes are low, the most significant concern is that the wells do not produce adequate quantity to be worth a significant financial investment in sensing units. Hence, the costs need to be lowered.
- *Remote Oil Wells Without Connectivity:* Oil wells are remote. They have no power or web connectivity and are tough to get to.
- *Pricey Hardware Installment:* Not just do the sensing units set you back a lot but also the installation likewise costs a whole lot adding to the price of the original sensor. Further, the software application needs to be constructed. For smaller procedures as well as even more remote end-of-life wells, quick ROI was the essence to justify the IIoT application.

The advent of IIoT innovation in the oil as well as gas industry makes it possible for operators to manage, analyze, and act on huge information collections from the many physical possessions deployed for the production and transport of hydrocarbons. The IIoT allows putting together fragmented information right into a structured data group, which can be assessed to determine patterns showing a possible risk or tools failure. The development of durable anticipating logical devices makes it possible for such problems to be dealt with by deploying positive steps, be it preventative maintenance or turning on very early caution signals. As a means to enhance

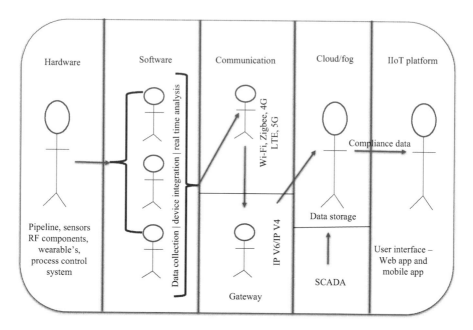

FIGURE 3.5 Function of IIoT in oil and gas industry.

business outcomes, oil and gas industry drivers take more interest in IoT modern technology for not only to enhance financial performance but also to preserve important uptime. Also, it requires more real-time, end-to-end connectivity of individuals, process, data, as well as points across the upstream, midstream, and downstream fields to improve functional performance and hence, return the service to development. That is why the fostering of IIoT systems along with Big Data are currently regarded as immediate concerns.

The applications of enterprise IoT in the oil as well as gas market are countless. Midstream market has the tough job of moving variable quantities and qualities of items from multiple areas to new end users as well as markets. For midstream service, IIoT plays a crucial function in keeping track of the following: connecting pipe networks, sensors, leak discovery, alarm systems, emergency situation closures to interact seamlessly, to be available for analysis and interpretation in genuine time would considerably reduce a few of the significant risks that this sector of the market handles.

Business Challenge

Under the progressive impact of product flows and vibrant environmental conditions, pipes are at risk to multiple structural failings. Rust, cracks, leakages, and deboning are among the most typical issues. The pipeline stress routinely change. The majority of pressure declines belongs to normal activities such as nuclear power plant drawing gas or chemical plants increase manufacturing. Other stress declines could be the result of strange conditions, such as leaks or ruptures. Regulatory authorities push pipe supervisors to manage risk posed by those fluctuations by constructing automated systems that can both precisely react to and detect abnormalities.

Precondition

Anxieties on piping parts occur mostly from inner stress, dead weight, and restrained thermal expansion. In view of the elevated temperature plus stress levels, incomplete relaxation is of significant relevance as are other feasible unfavorable influences, such as boosted friction in pipe supports, which swiftly utilize the staying small stress and anxiety reserves. Online surveillance of the levels of exhaustion of thick-walled components, the relaxing of the piping system over and above the capability of the hanger system, is thus an important precondition for maintaining functional dependability, developing maintenance along with evaluation periods, and the risk-free long-lasting operation of oil and gas plants.

Approach

Industrial wireless technologies can have a substantial result on oil and gas sector operations in two broad categories: wireless field networks for sensor and also gadget applications, and wireless plant networks for enterprise and procedure purposes. Wireless communication is among the most essential devices of transmission of information from one gadget to various other devices. In this modern technology, the information can be sent through air without requiring any kind of cables or other electronic conductors, by utilizing electromagnetic waves like infrared (IR), radio frequency (RF), radio-frequency identification (RFID), WLAN (Wi-Fi), satellite, and so on. Wireless sensing unit networks consist of numerous different types of IIoT-enabled sensors, covering a large variety of applications containing pressure, temperature level, circulation, and level or simply relaying a contact closure with a discrete transmitter.

The main parameters to be measured are as follows:

- Pipeline temperature
- Oil and gas flow
- Pipeline strain
- Leakage detection
- Deformation of pipeline structure

IIoT sensors mounted at regular intervals along the pipes help to keep track of stress, flow, compressor condition, temperature level, density as well as various other variables:

- *Acoustic sensing units:* Used to find a violation of variant in the acoustic signature.
- *Flow sensor:* Used to measure the velocity of the liquid.
- *Fiber optic sensors:* Used to find deformations in the pipe walls.
- *Piezoelectric stress sensors:* Used to measure changes in pressure.
- *Resistance temperature-level detector or a thermocouple:* Used to measure the change in the electrical resistance of a metal as a function of the temperature.

The wireless M2M communication options will deliver smart wellhead surveillance as well as control options to support optimal working, accurate telemetry, and fault-tolerant interactions. Versatile cordless M2M communications architecture delivers protected, end-to-end, high-speed wired as well as cordless communications for networks. SCADA systems can keep track of the pressure going down; together with IIoT system, they are ideally suited for stream evaluation of the data for real-time information based upon rules-based plans. If stress decreases and also the equivalent reasons cannot be presumed based upon readily available data, the system would certainly send ideal alerts.

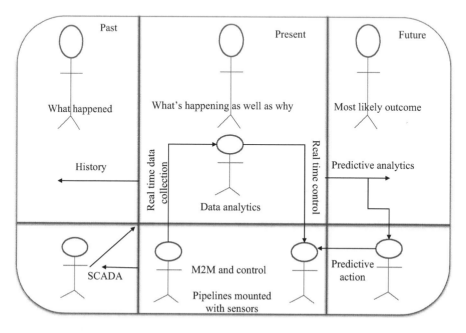

FIGURE 3.6   IIoT-enabled monitoring of pipeline in oil and gas industry.

Predictive Analytics

Wireless sensing unit networks contain different kinds of sensors, covering a vast variety of applications, including pressure, temperature, flow, and level or simply relaying a call closure with a discrete transmitter. Wireless sensor network applications monitor dynamically close to real-time process control, safety and security, governing, and manufacturing efficiency. The gauged information from the sensing units installed for each of these applications would certainly be revealed on an IIoT application using a web Internet browser and is obtainable from anywhere in the world by means of mobile application or desktop computer. With predictive asset monitoring, oil and gas enterprises get very early warning alerts of devices concerns and also potential failures, which aid them to take instant action and enhance general performance.

RESULT

The IIoT-enabled monitoring enables condition-based maintenance alerts. When the parameters in the pipeline drifts from its suggested specifications, the IIoT sensors proactively keep track of pipelines and send out an alert to the designated supervisors,

mining engineers, environmental engineers, and operators. This predictive maintenance keeps the life of pipeline running extra effectively and also avoid downtime.
Execution of IIoT solution for oil and gas industry assists to achieve its corporate
social responsibility that significantly decrease the effect of mining operations on
the environment. The power to improve productivity, control, as well as monitoring
the process with the IIoT-enabled monitoring and extensive information interaction
comes from the evolution of IIoT.

SUMMARY

Industries are demanding to improve systems and tools to fulfill technological
standards and policies, to stay on top of enhancing market to handle demanding
technologies. The collaboration of individuals, manufacturing applications, and
machines on the shop floor of enterprise forms the closed-loop environment. IIoT
matches both DCS and SCADA by widening existing capabilities like real-time info
capturing, machine breakdown signals, real-time control, data logging, information
evaluation, and visualization. The IIoT brings value to the manufacturing enterprises
without being constrained by economic as well as technical limitations. Industrial
automation in manufacturing uses smart machines in factories to carry out manufacturing process autonomously with minimal involvement of operators. With the
advancement of microcontroller, software program, automation in manufacturing
relies on the abilities of computer systems and software to automate, enhance, and
incorporate the various parts of the production system. The IIoT guarantees to use a
revolutionary, completely attached smart world with its relationship between things,
environment, and people. The limitless amount of connection available between
devices enables it to connect generally different aspects to become extra reliable and
customer-centric. Enterprises are swiftly uncovering new age of advancement in the
form of smart technology. Product manufacturers are moving toward making smart
products all over industries, which provide innovation and collaboration across the
supply chain. IIoT or smart manufacturing, the factory of the future, taking information from smart connected devices and allowing manufacturers to get insights into
real time will certainly be a lot more effective in the usage of man, machine, and
material. Smart connected products take on Industry 4.0 technologies, consisting of
sensing units, connectivity on top of mass customization, and smart materials. By
welcoming Industry 4.0, accelerative enterprises are utilizing smart modern technology and production specialists to make cutting-edge smart connected products meet
customer expectation.

BIBLIOGRAPHY

1. CIM data Webinar. June 25, 2015. Innovation in the Digital Age.
2. Ericsson. 2013. 5G Radio Access: Research and Vision, s.l.: Ericsson.
3. Ericsson-Slashes-IoT-Predictions. March 8, 2016.
4. Evans, D. April 2011. "The Internet of Things: How the Next Evolution of the Internet Is
Changing Everything." *CISCO White Paper Internet Business Solutions Group (IBSG)*,
San Jose, CA: Cisco Systems, Inc.

5. Usländer, T., and T. Batz. 2018. "Agile Service Engineering in the Industrial Internet of Things." *Future Internet* 10 (10): 100.
6. ISOC-IoT-Overview. October 2015, https://internetsociety.org/iot (December 15, 2015).
7. ThingWorx Platform|Product Brief|PTC, https://ptc.com/en/resources/iot/product-brief/thingworx-platform
8. Mikell, P.G. 1999. *Automation, Production System and Computer Integrated Manufacturing.* New Delhi, India: Prentice-Hall of India.
9. "The Intel IoT platform." https://theiotlearninginitiative.gitbooks.io/internetofthings101/content/documentation/Intel.html (May 23, 2016).

4 Smart Product Development

Modern technology is advancing at a rapid pace. Product development in the space of Industry 4.0 is more demanding, active, and comprehensive. Products are ending up being intricate and innovative based upon the client requirement in this global village. Taking care of the process from start to finish (NPD/NPI, process, launch, and service) is the crucial duty of an enterprise. Everything from devices to automobiles to handheld tools is connecting people to information and to solutions like never ever before. Developing smart products in this transforming environment is a difficult assignment. Smart product is a fundamental change that totally improves the exact result in a brand-new industrial revolution in NPD/NPI. To remain competitive, every manufacturing enterprise needs to work together with suppliers, OEMs, vendors and requires a unique collection of abilities that leverage technological growth of delightful product with seamless experience.

PHYSICAL TO SMART PRODUCT

Product managers specify consumer requirements through a structured technique and equate them into certain plans to create products. The ideal journey is easy, smooth, comfortable, and inexpensive. Although prospects are good for standard product development process, with current technological improvement, manufacturing enterprise needs to work closely with consumers moving toward even more accountable and cost-effective modes of manufacturing to face new realities. Progressive Industry 4.0 opens up connected world to competition, and customers are more demanding, in terms of smooth running and value-added services. This increase of a brand-new competition paves the path to alternative modes of product development process, which is trembling up the typical value chain and considerably gaining market share. This digital manufacturing revolution allows manufacturing enterprises to release their vision, maximize equipment dependability, production, improve top quality, and drive ingenious business models.

For a product development enterprise, these new challenges are all the more important because they are facing a paradigm shift. Any new entrant would immediately face imposing barriers to entry. Yet digital transition and market opening are breaking down these barriers. A network of smart connected product design and development is considered as a major determining factor. The growing awareness of the costs of product failures and the increasing complexity of industrial equipment are driving manufacturers to rethink their product development strategies.

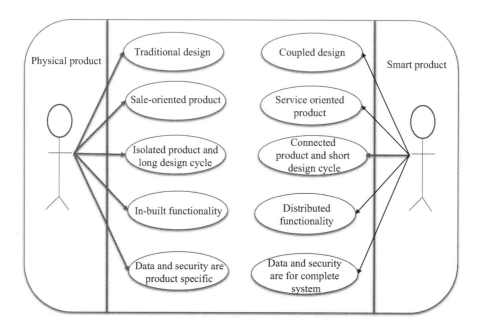

FIGURE 4.1   Transition of physical product to smart product.

Smart Product

The IIoT is changing the nature of physical things. It is helping to connect individuals with each other as well as provides information, thereby making it possible for any object or place to connect to the web. This connectivity, in turn, encourages challenge of providing details to people through a single source. The advancement of products into smart, connected gadgets is radically reshaping manufacturing firms and competitors. The connection between consumers and products is ending up being open-ended and frequent. In traditional manufacturing process, manufacturers make their products with integrated obsolescence to guarantee that the cycle of the product continued. Instead of products becoming obsolete and outdated, smart technological development now permits manufacturing enterprise to actually upgrade the products. Smart products substantially change the work of essentially every feature within the enterprise.

Smart products have a number of same physical components that the products actually constantly had with new attributes – sensing units and software application that make them much more intelligent. The three core aspects that create a smart product are as follows:

1. Physical components, including mechanical, electric, and electronics parts.
2. Smart components, including sensing units, microcontroller, embedded system, software application, cloud, mobile, and so on.
3. Communication components, including protocols, Wi-Fi connectivity, ZigBee, antenna, and so on.

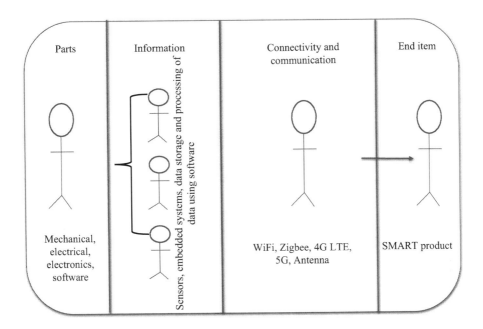

FIGURE 4.2 Components of smart product.

Data plays an important role in the product development activities. Previously, data were generated mainly by interior operations, details collected from surveys, research, and other outside sources. Currently, the sources of information are being supplemented by product itself. Data now stand on par with people, innovation, and resources as a core possession of the enterprise; their value raises significantly when they are incorporated with other data. Smart products make use of data collected, such as commercial properties like being context-aware, adaptive, self-organized, and positive along with having the capability to support the entire lifecycle.

Facility to view the amount of data ends up being a vital source of competitive advantage. It enables effective understanding by identifying patterns in thousands of analyses from many products in time and by utilizing organization analysis strategies. Accumulated data are usually disorganized emerging answers in data lake. The product data information collected from data lake is utilized and analyzed with data analytics tools. These capabilities include monitoring, standardization, controlling, and automation. Smart product innovation has the opportunity to truly adhere to the lifetime of its products as well as track their use throughout. Similarly, a product can in fact interact with the consumer, informing them the most effective means to make use of, look after, and also get rid of at the end of its life expectancy.

CONNECTED PRODUCT DEVELOPMENT

Processes utilized to design and build mechanical, electrical equipment usually starts with product research study followed by the development of concept through design, prototype, testing through manufacturing, and product obsolescence. Currently, in the age of Industry 4.0, the product manager is required to think about the business value chain for product development and launch, from smart digital viewpoint. In the smart connected world, manufacturers' strategic decisions in investments need to be made with the objective of becoming digital in every way led to a new approach by NPD team utilizing digital business models, processes, and tools. Management executives along with product manager describe the smart product that is utmost likely to be built and also the troubles they are likely to address through market requirement document (MRD). MRD in turn helps to produce a product requirement document (PRD). PRD will be needed by the NPD group members to establish the strategy for building and developing the product. MRD helps in specifying the critical investing top priorities and the operational areas most impacted, whereas PRD helps in defining the granular functionality of the product. Product managers are increasingly strategizing the use of complete digital perspective with smarter technologies and workable information, thereby generating far better insights into totally integrated digital services and consequently well positioned to be more competitive. The product data will flow bidirectionally between the processes, which permits faster understandings, better choices, cost reduction by shifting from physical to digital development, simulation, and testing.

Industrial product design is changing along with customer demands. The basis of product design has actually been to satisfy requirements of both customers and manufacturers. Including a layer of digital knowledge to the product reveals possibilities for new performance, far greater integrity, a lot greater product utilization, and abilities that crossed as well transcend conventional product horizons. Effective brand-new data and capabilities of smart connected products are reorganizing the traditional functions of business. Product development shifts from greatly mechanical design to true interdisciplinary systems engineering, that is, the design of the physical equipment including smart devices and smart products, need to cover the complete spectrum of technologies and skills across the cross-functional NPD team along with information technology and software programmers. Design teams are changing from a majority of mechanical engineers to numerous proficient disciplines of engineers.

Having quality view since the beginning of the product lifecycle all the way through production to distribution, manufacturers will be able to achieve higher degrees of customer satisfaction. Physical product design indicates developing an overall smart connected system. Industrial devices, for instance, utilize sensing units efficiently in producing considerable amount of data than in the traditional development process and send the collected information firmly for analysis followed by action. While exactly positioning the sensing units on the gadget, make sure the capability of the sensing units to operate in extreme conditions need to be factored while designing the product. Product advancement accepts advertising when it pertains to consumer experience problems and also both normally concentrate on form fit and functional requirements.

Product designer while designing the smart product will take into consideration the right modern technology for optimal ingrained knowledge, connectivity, edge component, cost, quality, and maintenance. The digital interface makes possible for remote process and even getting rid of the requirement for controls in the product itself. Augmented reality/virtual reality applications use the cloud that helps in

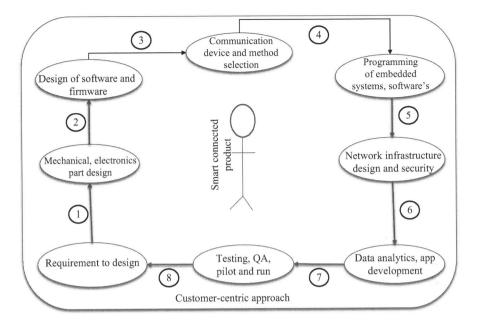

FIGURE 4.3 Smart connected product design.

creating a digital overlay of the product. A product that conforms to manufacturing standards presents no threat to the customer and is also considered to be of high quality. To make every effort for such high-quality product, businesses are required to develop a durable high-quality system, with integrated lifecycle monitoring to meet certain compliance requirements.

DESIGNING SMART CONNECTED PRODUCT

Industry 4.0 has been gradually changing the manufacturing industry into a revolutionary digital sector. Our everyday lives are full of smart connected products: from industrial, medical, wearable devices to durable goods; the products utilized in industrial sectors, home, and office are becoming so increasingly smart and connected that the world cannot be even envisioned without these products. The growing need for smart products across the globe is evident, everyone is engaged with evolving smart gadgets as well as smart connected factory. Customers wish to make brand-new cutting-edge products that are connected to the Internet, although they do not wish to sacrifice the bottom line with expensive smart integration. The NPD team strives hard to develop an excellent yet affordable smart product. Manufacturers have started incorporating advanced technological tools in design and operations. Smart connected things are changing in accordance with how product managers design, develop, and manufacture smart connected products. Smart products need to be created with quality and elegance in mind.

By designing smart products, IoT brings new revenue ventures for the product design and development enterprise. Smart connected systems gather data from physical tools and deliver actionable, operational understandings by integrating physical and digital components such as physical gadgets, sensors, communication devices, data hub, edge and cloud servers, data analytics, and dashboards. The key challenges of developing a smart device are economic limitations and intelligent design of product through significant variation. Smart product will certainly bring about enhanced overlapping of service coverage areas, thus sustaining unnecessary energy and financial losses. So it is important to make a cost-effective product, which is going to drive the smart connected infrastructure through different analyses: environmental analysis, service analysis, functional analysis, and user adoption analysis.

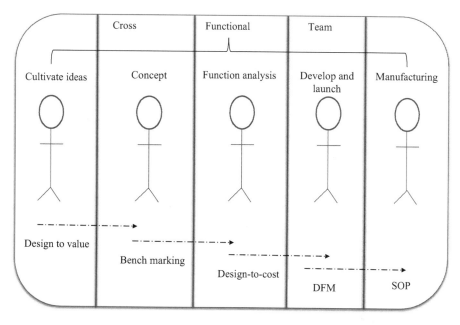

FIGURE 4.4 Designing smart connected product.

Reducing design engineering cost will/may lead to a substandard product layout and increase in material and labor costs. Reducing prices on parts might lead to greater service warranty costs and result in no transparency on client's understanding of the end-product. Adopting the design-to-cost (DTC) process helps product managers to use real-time supply chain information to evaluate design along with the expense drive in mind, such as manufacturing lead times, quantity rates, quantity capacity, logistics information, while the smart product is still in the design phase. With exact cost information required, the product development team will be able to take necessary action to prevent costly supply chain issues, such as parts availability or logistics issues, in real time. The NPD/NPI staff member and executive management have direct contact with customers to understand their true demands along with expenses directly to competitor's product pricing. The design engineer relates value with dependability, lines up with purchase individuals for the cost paid to them, and process and manufacturing engineers keep cost from the angle of manufacturing and sales people in line with what consumer wants to pay. A value analysis matrix is prepared to determine the cost for each feature by connecting the feature with a device or a component part of a product. The function analysis system technique (FAST) authorizes to go across cross-functional team members with different technical histories to properly communicate and fix issues that call for multidisciplinary considerations for establishing a smart connected product.

Launching a smart product to the market requires taking into consideration optimizing manufacturer's service workflow with the connected services, which focus on how much it may cost to build. So determining the cost involved in manufacturing is governed by few components such as embedded system, microcontroller, software, sensors, network infrastructure, smart apps, and so on. Whether constructing a brand-new product or updating an existing product, the design stage needs to be guided by the principles of customer requirement, material selection, connectivity mode, integration, and information collection. With this in mind, manufacturing enterprise requires to focus on how to exactly utilize the modern technologies in product design and manufacturing. The design for manufacturing (DFM) process figures out the product success in terms of both guaranteeing high quality, optimum production performance, reliability, and productivity and taking care of variables influencing product manufacturing: raw material, its physical and chemical features, and its accessibility for faster production. Product design budget directly influences end-product expenses. The DFM process helps in incorporating any type of design-related product to comply with product regulation efficiently, thereby reducing design cost. Industrial equipment, as an example, need to have sensors efficient in producing considerably extra information than in the past and send out the data firmly for analysis and decision making.

PHASES OF CONNECTED DEVELOPMENT PROCESS

The design and development of a smart product needs a much more intelligent approach. Before kicking off development, it is practical to originally validate the concept of a practical type of the end item, for which a prototype can prove to be of excellent help. An ideal option of PCB (printed circuit board), its layers, components, and also the awareness of a version adapted to the consumption requirements is the secret. The right selection of the communications network and the electric supply parts of the gadget, including the selection of batteries, adherence to mechanical design, interaction tools, software programs, interfacing with cloud, IIoT platform, and mobile user interface, are the crucial steps in the development process. Product managers, designers, developers, and IT engineers need to have a tendency to believe in terms of process and not in a collection of attributes, which continue in a particular order to accomplish a desired outcome.

Dexterity and iterative development based on normal responses and close observation are always worthwhile, especially as wrong decisions, misconceptions, undesirable growths, or modifications can constantly take place; however, they are quickly uncovered with dexterous techniques and can be resolved without shedding much time, money, and track record. A mutual understanding of client requirement, in par with the competitive atmosphere and constant techniques, and approaches enhances the product market share regulate the growth of new products.

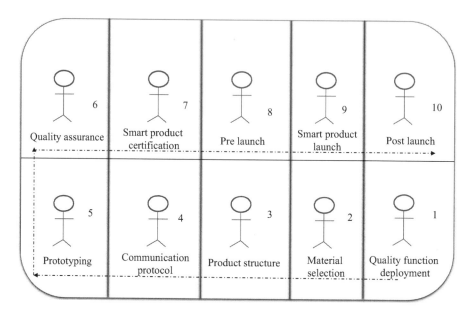

FIGURE 4.5   Stages of smart connected product development.

Quality Function Deployment

Environment-friendly smart products are becoming one of the most important concerns for the manufacturing sector forcing the manufacturers to boost awareness of the environmental factor to consider design and manufacture would meet the consumer and ecological needs. The principle of quality function deployment (QFD) is created to consider the quality of the product since the product design process is integrated with the ergonomics needs. Application of the QFD with regard to smart products has to be carried out extensively for the affordable product and services quality. It is the initial step in design that helps in converting what the customer desires into the product and services that satisfy the requirement with regard to engineering design values using house of quality matrix. It makes possible for the product design team to define consumer demands sustaining assessment of the product and services capability by refining data gathered to explain its impact on the requirement. Concept derived from the process will show the target audience for the product, its advantages, attributes, and connects in addition to the intended recommended proposed cost.

Material Selection

The innovative worth of a product is featured by the material involved; in fact, it is mechanical, electric, magnetic, optical, thermal, and chemical and physical properties, in addition to its forming, joining, as well as ending up technologies, that contribute to the success of a product. Smart material that forms the basis for developing a smart product describes the product ability of changing its properties instantly in reaction to an external inducement. To lessen product overheads, material engineers need to take into consideration the following: product design, product variant, choice of material, and manufacturing method.

Product Structure

Conceiving of the product concept helps product designers to provide smart product structure, that is, smart bill of material, and a visual representation with the help of heterogeneous CAD tools. The complete product lifecycle is almost altered at the design phase. Design of the smart product is based on the track record of the sensing units, printed circuit board assembly (PCBA), and so on. Smart products call for a brand-new strategy and also the growth of product structure called smart BOM. Smart BOM comprises a mix of mechanical, electronic, and software program components combined into products and becomes a vital component of the product design and development process with the ability of assisting manufacturers effectively launch a product to market. Complicated data frameworks need to be caught to create a BOM from multiple sources. Resources consist of MCAD, ECAD system and embedded software program system. The product structure of smart product integrates several sources of CAD data management, with a capability to keep an eye on modifications throughout every phase of the development.

Communication Protocol

The communication protocols of smart connected product occur with wired and wireless radio communication: Internet Protocol (IP), the communications process that supplies a recognition along with gadgets, computers on networks along with routes traffic throughout the Internet or intranet. Internet Protocol Version 6 (IPv6) is an important communication enabler for the smart connected product. Smart product calls for a mode for sending the details noticed at the gadget to customer for subsequent decision making. Depending on the need, short-range communications such as Bluetooth, ZigBee, RFID, near-field communication, antennas, GPS, 2G/3G/4G/5G, and so on are typically made use of together with wireless LAN or wide area network. Choose the best communication protocol considering the existing network infrastructure.

Prototyping

Prototyping of the smart product is the first handover to identify the best path to make the customers satisfied. It is the process of integrating hardware devices to software enhanced with smart sensing units, PCBA, embedded systems, and microcontrollers. Evaluate the smart prototype product completely: materials, communication devices, comfort of use, and the cost involved. Once the budget plan is approximated, start building your first smart product. The prototype product is made use of to understand the important affecting factors and frame out the necessary parameters of smart product deployment. Development process will get smoother when the problem is solved during the prototyping phase in collaboration with the product development team. Proto model must be end-to-end, linking the sensor through the device, network, cloud, end-user interface, and business integration. Full-scale development of the smart product skipping the prototype phase will blow enterprise money. Developing a smart product will have the customer involvement from initial phase through the complete process, delivering true business value to the customer.

Quality Assurance

Because of the quick development in connected technology, manufacturing enterprises must ensure that the smart products are constantly and also completely evaluated prior to being carried out in this connected world. Testing of the smart product needs to be tightly incorporated right into the product development process as it is crucial for complicated combinations and changes; therefore, organizing dedicated quality assurance with testing features with a mix of systematized and decentralized approaches is essential. Security and efficiency testing is a top most priority of the business. One of the feasible options is to build a smart connected lab that offer the entire digital portfolio to experiment and also imitate dynamic experiences that would certainly assist smarter means of testing comprising usability, compatibility, combination, protection, and efficiency testing.

Smart Product Certification

Smart products are equipped with Internet-connected sensing units in addition to several radios that create and also send out data to its main system. Such real-time data handling requires a particular quantity of power from the device. Smart connected tools with wireless connectivity and/or battery capacity are subject to related policies to ensure consumer safety. The products that traditionally were not wireless in nature are getting transformed to being smart. There are several kinds of certificates that smart product require to be used commercially, including Federal Communications Commission (FCC) certification, which is required for all smart connected product offered in the United States, significantly by more expensive smart products using wireless connectivity. UL/CSA certification for smart product is necessary in both Canada and the United States. European Conformity (CE) certification is the European Union analogue of FCC and UL. RoHS accreditation is a standard certification across the globe which validates that a product does not have lead.

FCC/CE/UL certification screening ensures that new product designs fulfill the defined CE/UL/FCC norms; similarly, in some scenarios, LoRaWAN certification is also applicable. Product development enterprises need to get their smart connected product cleared for aforementioned compliance screening to get them to market at the earliest. General emissions testing and intentional radiation testing are the main testing required for the smart product that is making use of a cordless technology such as Wi-Fi, Bluetooth, or ZigBee to get certified. Other safety and security compliances need to be taken care of: designing the smart product should adhere to compliance of restriction of hazardous substances and energy star rating; communication technology such as Wi-Fi, Bluetooth, cellular 2G/3G/4G/5G used by the smart product should adhere to the electromagnetic radiation direct exposure standards.

Prelaunch or Pilot Run

Prelaunch or pilot run involves validating interoperability of heterogeneous smart products called for to complete a service for consumers, ensuring smart products are being verified in real-life scenarios by leveraging prelaunch examination groups.

Cloud-based distributed examination framework helps in remote accessibility to crowd-sourced smart connected product in a secure way, allowing product development group to confirm the smart product compatible with other heterogeneous system, by utilizing the application mimicking customer real-life test situations and offering feedback on the system for taking immediate decision during the product development phase before moving for a final production. Pilot represents lower danger, but it permits rolling back to the previous capacity of doing things if required. These remarks come in handy, making the application durable enough for the production deployment. Users usually have an expedite system for getting assistance from the product development group.

Product Launch

The secret to effective launch of smart product begins with clear goals as well as interacting to guarantee those broader goals are fulfilled. Smart connected product open doors to brand-new markets with new service designs. Releasing a smart product provides major difficulties, although with the best capability, team, and buy-in from all stakeholders and customers, manufacturing enterprises are well-positioned for success.

Postlaunch

On introduction of the final end-product to the customer, a smart product ends up being an active thing to start communicating within real time. The best practice to stay long in the market is to learn more from customer comments than what found during the testing stage of product development. Have a detailed road map of needs to show the product direction and the impact of new features added to the smart product make good sense for all stakeholders including clients.

Moving out a successful smart product is not a very easy accomplishment. One of the main features of the smart product development from voice of the customer to product launch is predictive maintenance, which is extremely important to the consumer expectation as a product development and manufacturing enterprise constantly faces obstacles when interacting with the product vision throughout business. It is an uphill struggle, yet it is likely the most crucial function of the product service. The result is increased openness, which results in far better communication, delighted teams, and the most important satisfied consumers.

SMART PRODUCT ECOSYSTEM

True value depends on industrial control systems that are comprised of smart connected product, equipment, and devices that function as a smart connected ecosystem. Enterprise value increases as the data result in greater orders of actionable material, which moves from straightforward detailed states to diagnostic, anticipating, authoritative states that enable real process optimization to inevitably providing a semantic type of data considering social environment at large. Designing of connected products combine physical together with digital parts, collect information from physical devices, and deliver workable, functional insights. These parts consist of physical tools, sensors, information extraction and secured interaction, gateways, cloud servers, analytics, and control panels.

Smart product ecosystem adheres to the following criteria:

- Compatibility is the underlying principle in all smart product design and development processes.
- Data visibility, wherein the physical processes are recorded and saved generating a digital twin.
- Industrial assistance is the main ability of smart connected systems to offer and present data that assist individuals to make far better functional choices and also solve problems much faster.
- Smart connected systems outdo assisting and interchange data, which helps decision-makers to make decisions and also implement requirements according to its defined reasoning.

Creating a smart connected ecosystem ought to effectively fill its objective (problem solving, digital twin, predictive analytics and maintenance, quality administration, supply chain optimization) paving path to create enterprise value. With the advent of digital twin, information is gathered constantly from sensors and is passed to the physical counterpart throughout the system's lifecycle. Engineers, designers, software programmers develop digital prototypes, then run simulations to check the product's functionality using fog computing, as every information from manufacturing procedures to sensor input, to external management software application can be fed right into, and arranged within the digital twin, it is easy for designers to repair noted errors in the product.

As smart connected ecosystem develops, consumer devices will certainly have to ship with the right networking and also automation functions. Cloud computing, IIoT platforms, and artificial intelligence combined with machine learning help in linking networked devices, handle applications, process information, provide reports, and include sophisticated analytics functionalities.

SUMMARY

Enterprise starting the physical layout of the connected smart product, whether by retrofitting existing products or developing brand-new ones, there are a host of elements they require to account for initially. Physical layout indicates even more than creating connected product, and developing a general smart system. Devices and products look the very same as they did before they obtained connected and smart product. Enterprises redefining their solution or product experiences for smart connected product are typically not able to see beyond the intricacy. Low-cost design is the rival for aggressive time to market strategy. Develop a smart product so as to utilize an off-the-shelf service such as a reference design, and change it as low as feasible. With a dedicated team and also the right tools, product development team can reduce product prices and improve the performance of the smart product. Quality tracking permits product managers and manufacturers to constantly improve product quality over time, while the efficiency information gathered from product or production equipment supplies an indispensable source of information and product development insights. The end-to-end customer experience optimization, operational flexibility, and technology are vital drivers and also the goals of smart product development, in addition to the generation of new income resources with connected lean ecosystems of value, leading to enterprise business improvement. The secret success of this journey starts with several connected intermediary objectives relocating toward continuous optimization throughout processes across the smart lifecycle. The execution of data-driven service and also service designs for smart connected product development, ingenious methods, cultural change, and an inspired group are required, because its growth rarely stop working because of technology or the budget plan. When they stop working, it is a result of the existing company culture, absence of flexibility, silo structures, wrong process, and lack of expertise. Smart connected product made a huge impact on traditional product design, which undergoes a shift from being product-centric to experience-centric. To be successful with smart product development, the limitation and high-risk factor is not the technology but the approach. Recognizing the production volume and evaluating the time-to-market effect for a smart product will help manufacturers to stabilize the price from an enterprise point of view.

BIBLIOGRAPHY

1. "New Product Development Software – Portfolio Management|Planview." 2019. Planview. https://www.planview.com/products-solutions/products/ppm-pro/npd-portfolio-management/.
2. Ward, A.C. 2007. *Lean Product and Process Development*. Cambridge, UK: Lean Enterprise Institute.
3. Ibrahim, Z. 2000. *CAD/CAM Theory and Practice*. New York: Tata McGraw-Hill.
4. Kenneth, B.K. 2013. *The PDMA Handbook of New Product Development*. Hoboken, NJ: John Wiley & Sons.
5. Bolton, W. 2010. *Mechatronics: Electronic Control Systems in Mechanical and Electrical Engineering*. Harlow, UK: Pearson Education.
6. Ong, S.K., A.Y.C. Nee, and Q.L. Xu. 2008. *Design Reuse in Product Development Modeling, Analysis and Optimization*. Hackensack, NJ: World Scientific.
7. "Product Lifecycle Management for Smart, Connected Products|PTC." 2019. Ptc. Com. https://www.ptc.com/en/products/plm/for-smart-connected-products.
8. Timoshenko, S.P. 2004. *Strength of Materials*. New Delhi, India: CBS Publishers & Distributors.

5 Convergence of PLM with IIoT

Connecting to the Internet can be accomplished with a range of communication innovations. At the heart of most Internet of Things (IoT) and Industrial Internet of Things (IIoT) products are one or more wireless connections that enable communication with the World Wide Web. The biggest wave on this smart globe is washing through innovative manufacturing (digitization), typically a much better technique for sharing all information during the whole product lifecycle, leading to lower prices, faster turnaround, and higher quality. Intelligence embedded into the product in the type of sensors along with software application is the core element of IIoT, contributing context, feature, and evaluation of the raw data collected. Manufacturers are constantly concentrating on high quality, performance, stability, and favorable connection with customers. To utilize information coming back from IIoT, PLM requires brand-new capacities, including both connections, to take in sensing unit data, resulting evaluations, and interpretations in order to link that information with the cross-function stakeholders of NPD/NPI driving improvements across the product lifecycle completing an end-to-end connected loop. The value of PLM to IIoT cannot be downplayed; IIoT will basically change the means by which companies develop products, its growing demand and new technologies will be well integrated with PLM in order to be successful. It has really become evident that adhering to generation of competitors is automatic and appears to need change in everything, from making products to offering continuous tracking of service. The IIoT information insights complement PLM, stimulating product technology, which triggers a great deal of much more polished product understandings to different level.

DIGITAL COLLABORATION

Digital collaboration extends PLM to include process engineering, which consists of process planning and features within the product engineering, as well as other components of the lifecycle process. It facilitates the holistic view of product and process design as integral parts of the product lifecycle that make it possible for product design to be conscious about process constraints and capabilities. Digital collaboration attaches the product data that are generated at every lifecycle stage: evaluating client demands via delivery, post distribution upkeep, and servicing of the product. The product lifecycle process is directly linked with product quality, relying on predictability to attain the desired purposes. Integrating manufacturing process quality systems with PLM allows enterprise to maintain quality records adhering to regulative needs.

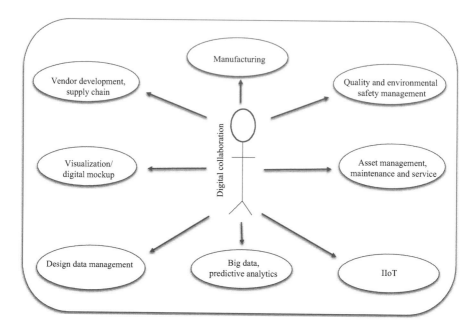

FIGURE 5.1   Digital collaborative environment.

The product development team requires to develop cross-functional design, firmly related to information technology, calling for a greater level of collaboration and also data integration. The product data is available once they have manually become part of a PLM/PDM system. From there, they are incorporated with various other manufacturing/shop floor applications to develop a digital collaborative topography. The process in a PLM chain might be swarming with flaws intimated through problem report, and rectification takes weeks to compile and implement through change process. It seems easy to see exactly how IIoT can be used to involve consumers, but the strategy lies in just how to connect these collected data back to all the team members of a NPD/NPI.

The PLC, SCADA, sensors, and smart devices progress to become a lot more inexpensive and also efficient, paving path to track real-time performance of the products by the customer in the day-to-day activities. The NPD/NPI groups work together digitally to develop products in which iterative changes or improvement is recognized and integrated into the process. The NPD/NPI group guarantees quality and compliance with the sensors, thereby creating a much more efficient, streamlined technique that allows product development and business stakeholders to utilize new information and understandings, delivering lasting enterprise value to customers. Product data deliver precise, updated, actionable details to all who need it, and also coordinate their responses to it by digital collaboration. Expanding this string into the operational phase provides deep understanding of field operating problems and the product experience is an essential involvement of the product digital twin.

The IIoT enables huge quantities of information be readily available to quality participants in real time, and also information from several resources at the same time will be made use of to enable fast, situational decision making. Product quality management adds benefits to the enterprise in the form of high-quality process to be an ingrained, important component of the ecological community. Enhancing quality will certainly require to accept information and utilize innovation seamless to drive collaborative digital product and process development.

Digital Collaboration in Product Design and Development

Collaboration is the most challenging part of the product design and development process. Unlocking the ingenious potential throughout the NPD group offers an inspiration to development by involving all stakeholders and cross-functional team from starting in a workshop-based strategy, thereby developing a feeling of common possession in the success of introducing the product to market. In the very early days, hand-drawn to CAD-based design product managers in manufacturing industry have been deliberately concerned with the pressing opportunities of design innovation: starting with paper illustrations and gradually transforming to digital drawing through digital mockup. It opens up stage to concurrent engineering, involving methods, processes, and tools to serve all functional designs. Today in the modern digitalized age, integration of functional and industrial design teams is possible by digital collaborate engineering.

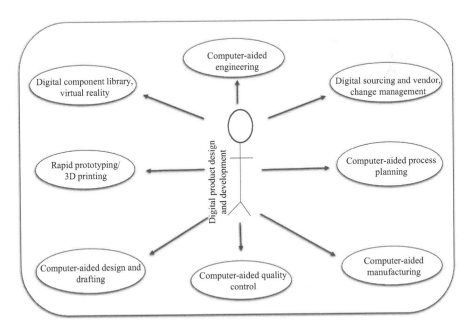

FIGURE 5.2   Digital product design and development.

The successor to CAD is the digital twin, which is being used by engineering and processing teams to describe information about the physical product and its behavior in the real world with the digital representation and other areas of the business accelerating the convergence between the real and virtual worlds. Manufacturing enterprises make use of digital twin technology to assess the efficiency of physical assets, thus paving the path to renovations to get better end results. It helps product design and development to develop different concepts, and performance of the products is tested virtually, thereby helping in the improvement of product in the product development process. PLM forms the foundation for digital twin.

The collaboration technology of PLM eliminates manual process associated with design and development process, by the coordination of the product data flow across the NPD team digitally. The digital collaboration in product design and development makes it easy for the NPD team members across various departments to collaborate digitally on PLM items, such as Parts, BOMs, CAD files, ECOs, projects supplier, vendor data, process data, and other data related to product development.

VALUE DRIVE OF DIGITAL COLLABORATION

Concept-to-product design is generally focused on the advancement of a product, whereas process engineering is mostly concerned with the advancement of the process to manufacture the product. For a long time, the manufacturing industry has been engaged in the renovation of digitization. The journey starts with CIM (computer-integrated manufacturing), ERP, MES, production planning and scheduling, supply chain management, quality management, risk management (failure modes and effect analysis), design for manufacturing/assembly (DFM/DFA), design for environment (DFE), design for testing (DFT), PLM, and so on. Thus, manufacturing industry has begun to transform through increased use of digital technologies.

Slowly, PLM has started to redefine the value of data in the manufacturing sector, transforming the product development in today's digital age. Regulatory compliance and data traceability require an effective integration of design and manufacturing applications. PLM is considered the backbone that allows successful collaboration with other manufacturing applications. It started with integration of PDM with PLM followed by ERP with PLM. The development of technologies such as computer-aided design, computer-integrated manufacturing, computer-aided engineering, and digital mockups forms the basis of digital collaboration.

Manufacturing industries are recognizing digital collaboration ahead, with PLM as a resource around product development and product design data becomes a key to feed the path to smart manufacturing. To reap long-term benefits of PLM through Industry 4.0, integration between PLM and other industrial assets enables design and manufacturing divisions to exchange product data information between digital design and physical manufacturing execution. The starting point for the path to Industry 4.0 to support a digital enterprise involves mockup, visualization, additive manufacturing, augmented reality, virtual reality, analytics, asset management, robotics, inventory monitoring, and quality management that enhances collaboration between product designers, customers, manufacturing units, and harmonizes various processes with the advent of secured IT infrastructure, which are the milestones of this journey.

The evolution of smart product is revolutionizing product lifecycle management. Several trends and technologies are disrupting PLM such as Big Data, mobility, IIoT, artificial intelligence (AI), and machine learning along with the involvement of electronics and software BOM in manufacturing. Customer demand for integration in product design and development with operation process opens a world of interactive experience, capable of capturing the product requirements, predictive maintenance, and services, thus providing manufacturing enterprise an end-to-end lifecycle management.

PRACTICE OF ACCLIMATING DIGITAL COLLABORATION

Manufacturing enterprises are in the new age of industrial change, a digital collaboration journey controlled to impact all aspects of business from the enterprise structure to how they generate revenue by the utilization of digital data, connectivity, and process covering every facet of manufacturing activity in the shop floor. Digital association among product managers, designers, process engineers, operators, customers, and physical industrial assets will certainly open enormous worth and alter the enterprise digital collaborative landscape. Product and process information is continuously being shared with supply partners and the distribution chain, by real-time integration with large number of systems and devices. The process of adapting digital collaboration is achieved by digitalizing physical assets throughout the value chain creating an end-to-end digital ecosystem. IIoT platform provides the integration and coordination required for increase in operational efficiency and digital collaboration of the business in driving technological returns.

The first step in fostering of digital collaboration in manufacturing enterprise is the financial investment in IT framework and automation of service process. Manufacturing enterprises demand to use digital initiatives to digitizing their process, not just to fulfill customer's expectations but exceed them. Computer-integrated manufacturing, with the assistance of digital devices to automate manufacturing process, is interfering with product markets. The trick to be successful for the manufacturers in this period is clearly to have a detailed electronic collaboration method in place to begin. One more big challenge is the change management that this activity stimulates, disrupting heritage systems to generate the brand-new adjustment innovations, which calls for transforming frame of minds, holding workers hands, coaxing lots of retraining, and urging those that reveal interest, as they can be the new adjustment agents.

Interaction along with data-sharing by generating a breath of fresh air with regard to reasoning enables space for strong technologies and restructuring the old system. Digital collaborative approach is released in a technique, detailed, and regulated, thus providing the manufacturing enterprise sufficient time to educate cross-function team members and also prepare customers for brand-new offerings. Actually, these are all positives, which are appropriately welcomed by the forward-looking services.

Few manufacturing enterprises do not have the appropriate digital transformation strategies, for example, ERP, PLM, MES system, and have a traditional application that is archaic. Likewise, the issues faced on the cybersecurity front have not included innovations like AI, Big Data, analytics, business intelligence (BI), and IIoT to their collection. All technological improvements equally ripen the foundation of a digital collaborative strategy.

Enterprises recognize the relevance of supply chain monitoring in their digital method; preserving the supply chain openness is an outright requirement considering the major effects its lack may have. To remain affordable, manufacturers should take advantage of the digital cooperation, making use of a mix of various devices, modern technologies, and techniques. To start with, they require to implement an ERP system adhered to by PLM system and improve their existing IT infrastructure. Customers desire problems to be fixed much faster, products supplied quicker, and inquiries addressed immediately. Digital collaboration enables business to alter the core idea and methods, making it feasible to increase over the extreme competitors in the industry.

DIGITAL COLLABORATION TO DIGITAL TRANSFORMATION

Innovation has actually encouraged customers, resulting in improvement in manufacturing sector, with enterprises adjusting to continue to obtain what they desire, whenever they desire, as well as exactly how they desire it. Collaboration between manufacturing business stakeholders and the IT team is critical for building confidence and overcoming challenges in the journey of digital transformation within an enterprise. Throughout enterprises digital collaboration between functions and services are breaking down silos, connecting all in an effective way. Business stakeholders prioritize revenue growth and reduction in costs, while IT teams prioritize integration within existing systems and overall security.

To remain competitive, manufacturing enterprises have undergone a shift from digital collaboration to digital transformation. The use of electronic product data, connection, incorporates from research and development, prototype, and testing, through operational efficiency, digital transformation affects all aspects of a product design, development, manufacturing, and service task. It leaves a tremendous influence on the enterprise available for constantly transforming in-house factors, competitors, and latest technologies. Digital collaboration to digital transformation supports in renovating the future of manufacturing by improving process, conserving time and money, and generating closer associations with customers in a closed-loop environment.

Digital modern technologies change manufacturing industry into entirely brand-new means, by providing essential enhancements in customization, performance, functioning, safety, and security. Digital collaboration is the vital stimulant of digital transformation program success. Efficient product data management is the first important step toward enhanced cooperation between service lines and features while enhancing compliance. Start of the digital journey try to find methods to enhance and to benefit return on investment (ROI), through detailed unified, distributed interactive and collaborative enterprises. Digital collaboration and transformation complement each another. Digital collaboration serves as a driver to move digital transformation initiatives ahead, bringing in new business and methods of collaborating; each of them cannot exist without the other.

INDUSTRY 4.0 LEVERAGE PLM

Industry 4.0 is a technical modification of this era. Without a system in position to provide top-level, comprehensive understanding into what is taking place, it can be challenging or difficult to make big picture selections effectively and efficiently within the enterprise. So, exactly just how it is viable? One option: "Industrial Internet of Things." Product lifecycle management that aids in dealing with a lot more complex supply chains will take place with IIoT by connecting straight to the connected suppliers; it also helps to ease access at an early phase in the product development process in order to conserve resource and time.

Prospective for IIoT options is substantial; numerous businesses discover that reaping the incentives is much easier said than done. There are facets of PLM that stand to dramatically impede the progression that IIoT can make within the commercial sector. To fully realize the potential of IIoT and to utilize product data successfully, products have to comprehend the difficulties brought about by PLM and take the required actions to address them. Product information is beneficial, which is used to drive necessary and reliable change. Taking the first action toward making use of information in the best means calls for having accessibility to it in the initial stages.

For manufacturing industry, variable gadget data are usually restricted within the engineering process by PLM. The longer the data are stuck within the PLM facilities, the longer they will certainly take to find raised value from the information accumulated about product and services. It is only when the product data are released, manufacturing industry has the ability to experience its real potential. With the power of even more data-driven decisions, manufacturers can take preventative actions when it comes to the upkeep and renovation of their assets. Enterprises often make use of a solitary spread sheet to distribute product information when needed, the solution is both short-term and high-risk. Also, the development with advancement of IIoT will soon advance beyond the capacity for giant spread sheets of data to effectively save as well share product information. Yet manufacturers are not the only ones with useful data to share. End consumers may additionally have information that can help to improve PLM preparation and also drive functional effectiveness. The data gathered from smart devices, consumers, suppliers, and also multiple divisions internal to a manufacturing enterprise leverage PLM process to develop enhanced wangled product.

SMART PLM OR I-PLM 4.0

Enterprises need to consider exactly how to successfully resolve the problems of data ownership and security: attending to the challenges of data sharing, ensuring the interoperability of different systems, and above all boosting the interaction networks in between all machines as well as sensors. PLM never ever actually relocated previous product advancement because it was not able to catch information after product had been delivered to customers, which would be important for areas like field procurement, support, or marketing. IIoT information coming back into PLM systems enables all product-related information to be offered for decision-making by the NPD/NPI teams, not simply product development or design.

Time is transforming numerous patterns and technologies interfering with the focus of PLM process like product lifecycle data, workflow, and IIoT in addition to the development of electronics BOM and software BOM in manufacturing. The growing demand for assimilation of product design and development with operation bring about a brand new form of interactive experience. PLM provides an interaction system to record concerns, comments, discussions from internal sources, consumers, vendors, and machines. The placement of PLM with enterprise product innovation objectives and manufacturing operational approaches will allow a change, enabling PLM as a core system to support product developers capturing the product demands along with anticipating upkeep and services. Product tracking can automatically link client responses to product records using ECO equate to quality product.

The quality of the product comes with a price and is frequently a considerable contributor to the total expense of top quality incurred by the manufacturing enterprise. As a best practice determine to test the quality of the process rather than the final end-product, achieved with the introduction of IIoT that helps in monitoring the manufacturing process to ensure the process is error proof. Valuable business choices are made when real-time sensor information is evaluated and immediately acted on, assuring quality near to six sigma level will be maintained.

Industry 4.0 paired with PLM uses observations where a part of the output is fed back to the input to improve and also lower errors of security of the product. The capacity to record real-time information from products in the field opens up brand-new use instances for PLM systems past the typical engineering-centered workflows of managing change orders and also handling costs of materials. Such massive cooperation can lead to Smart PLM or I-PLM 4.0.

I-PLM 4.0 leverage the worth from smart connected properties and machine-to-machine communication (M2M), such as equipment, devices, and machines on the shop floor. Real-time information from physical assets is incorporated with various other systems, such as ERP, CRM, and data warehouse, to supply comprehensive understanding of the manufacturing flow and also high quality.

BUSINESS CHALLENGES OF DIGITAL TRANSFORMATION

Getting any type of modern technology for the function of enterprise renovation includes some degree of forewarning. Technology amendment with customer engagements is a crucial element of digital transformation within an enterprise. Employing digital application along with existing application within the enterprise assists in the journey toward an electronically encouraged future. Staying up to date with the technological updates to compete when it pertains to digital transformation, there are various other variables that need to be taken into consideration.

Lot of new technologies and different processes across different departments with different members are involved in the product design through operations to customize the process as per the business need, with the digital transformation likely to damage down at some point. A primary concern in business is to comprehend the difficulties developing consequently and how to resolve them. Enterprises ought to look out ways to advertise the benefits of transformation to all department stakeholders, establish clear expectations, and provide training to equip themselves with the updated digital environment. So, what are the challenges?

- Roll-out of digital innovations comes to be a bigger challenge within the enterprise with different processes and cultures spread globally.
- Change is the most critical challenge, as most business stakeholders are reluctant to adhere; this results in performance and employee efficiency drop during the transition phase.
- Business and technology are uncertain, so proper connectivity between the two will enable to transform business operations and adopt an ongoing culture of innovation.
- Poor planning with an absence of sources leads to lower the performance of business.
- In receipt of familiar with new data monitoring techniques, training of staff members due to integration of new business applications.
- Having the right framework developed to scale the modern applications, so that the enterprise's digital structure is scalable without endangering data safety and security.
- Acceptance needs to have buy-in from every stakeholder for proper functioning of the automated business process.

MITIGATION OF DIGITAL TRANSFORMATION CHALLENGES

Identify the current digital environment to start off the transformation journey. Developing awareness of the importance of digital ecosystem to workers, customers, and stakeholders of the business as well as the relevance of change for consistent efficiency of product development and manufacturing operations. Digital innovation has to not only accomplish the certain preferred enhancements within the confines of the business process in a department but also interface, support, and help with the total requirements of various other divisions and the enterprise overall. A great location to start is with mapping of customer journey gathering consumer information via monitoring, study, and meetings. The digital journey mapped are implied to help with reasoning and overall business strategy.

The success of a digital transformation counts on the people executing the solution along with modern technology. To attain digital transformation success, manufacturing enterprise need to have detailed strategic planning, lasting approach, skillful resources, and quality of budget. Technologies such as social, mobile, analytics, cloud, and IoT are the essential aspects of digital transformation, which opens up more communications as well as possibilities. Digital ideation requires cooperation to attach people with other resources in manufacturing enterprise, beyond practical silos, with a digital thread for the product development.

BENEFITS OF DIGITAL TRANSFORMATION

Manufacturing enterprise, prepared for the succeeding industrial revolution, needs to transform electronically to get rid of the intricacies and open brand-new potentials. Customers want beneficial solutions to the problems. The primary focus of digital transformation is to use cutting-edge modern technology to boost the client experience. Data analytics, social media, apps, and smart devices revolve around a desire for a less complicated means of manufacturing a product, from the concept to the reality. PLM incorporated with ERP and integrated with MES and other legacy application enables enterprise to capture, digitize, and access all activity smartly on the production line, which helps suppliers to find out dynamically what is occurring constantly. Enterprises rebuild their business process with cutting-edge modern technology to bring the vision come alive and be successful in the smart connected future. To take complete benefit, businesses have to buy in new cultural change, standardization of process, training, and equipment. Operational excellence is an important key to differentiating their product service from competitors and keeping clients pleased in the smart connected globe.

Manufacturing operation is streamlined with minimal downtime as a result of connected makers sending out important maintenance information that can be leveraged to avoid malfunctions and maximize result. Making use of data insights helps to recognize customers and feeds into business strategy enables real-time feedback. Successful digital transformation has the ability to track metrics, analyze the data that are acquired, and utilizing the understandings permits companies to maximize enterprise techniques and processes. Product design and development processes are much faster and better notified using devices, such as additive manufacturing, virtual/augmented reality and leveraging behavioral information from users dynamically. Digital transformation replaces the legacy business processes with automated process and produces new service models to boost the effectiveness, resulting in better outcomes to the particular demands of the clients.

Sensors embedded with the machines in the shop floor keep track of production criteria of the product along the entire production line. Quality resultant from production data helps to analyze the source of problems and predict waste-related issues before they occur. Manufacturing enterprises enhance the effectiveness and make enhanced decisions by streamlining the entire work process by adopting collaboration, automation, usability, and learning from the environment around them, which aids to make recommendations for process improvement. Digital innovations are producing a breakthrough opportunity for business to enhance performance and improve quality. Integrating lean principles into digital transformation can be a highly efficient means of attaining extreme simplification of the process. It permits manufacturing enterprise to determine the most reliable controls for the digital journey.

SUMMARY

Industries are demanding improvement of systems and tools to fulfill technological standards and policies in order to stay on top of enhancing market to handle demanding technologies. Collaboration of individuals, manufacturing applications, and machines on the shop floor to enterprise forms the closed-loop environment. IIoT brings value to the manufacturing enterprises without being constrained by economic as well as technical limitations. Industrial automation in manufacturing uses smart machines in factories to carry out manufacturing process autonomously with minimal involvement of operators. Suppliers and customers are emerging as crucial components in the PLM ecological community. While product managers are providing product design ideas, consumers are leveraging social media networks to give comments bordering the look and feel, capability, and functions of the product. In fact, social media and also analytics are progressively tethering the standard engineering workgroup to the front end of innovation and pressing manufacturing procedures to the downstream to prevent high-quality escalations. To harmonize sustainability and compliance, manufacturers will require to adopt a step-by-step method to welcoming connected Smart PLM or I-PLM 4.0. The incremental adjustments may mean attaching product growth needs with their existing PLM framework and it might be everything about connecting to the real-world efficiency understandings on their smart connected products. With the advancement of microcontrollers and software programs, automation in manufacturing relies on the abilities of computer systems and software to automate, enhance, and incorporate the various parts of the production system. With IIoT or smart manufacturing, the factory of the future takes information from smart connected devices and allows manufacturers to get insights into real time; this will certainly be a lot more effective in the usage of man, machine, and material. For the essential demand for absolutely effective PLM, digital journey through IIoT is the end-to-end digitization of process, process automation, and coordinated human beings, which helps in leveraging the value chain of smart product management to smart service management and thereby influencing consumer satisfaction and reliability.

BIBLIOGRAPHY

1. Li, W.D., S.K. Ong, A.Y.C. Nee, and C.A. McMahon. 2007. *Collaborative Product Design and Manufacturing Methodologies and Applications.* London, UK: Springer.
2. Wang, X., and J.J.H. Tsai. 2011. *Collaborative Design in Virtual Environments.* Dordrecht, the Netherlands: Springer.
3. PLM | Product Brief | PTC, https://www.ptc.com/en/products/plm
4. PLM | Product Brief | Siemens PLM, https://www.plm.automation.siemens.com/global/en/products/plm-components/

6 Industry 4.0 Technologies That Enhance PLM

The concept of product technology and development is no longer immune to digital forces. The change starts at the base of the value chain and proceeds throughout the product life process. PLM develops and manages the final end product, defining exactly how each product was made and built to meet customer expectations, including expected operating problems, individual habits, performance results, quality forecasts, and service. Change starts at the base of the value chain and proceeds throughout the product life process. Manufacturers will certainly be needed to maintain the high quality of their products for a vibrant target market. Technology-driven organization strategies will benefit earnings in addition to price. PLM cannot satisfy the growing requirements of customers alone. Manufacturing enterprises are making substantial regulations in product design, BOM management, product compliance, simulation, and validation. Current PLM systems are trying to balance a blend of PDM and CAx devices with Industry 4.0 innovations consisting of analytics for handling Big Data and advancement of "Smart Product to Smart Manufacturing" by integrating Industrial Internet of Things (IIoT), embedded with sensors and software program that make possible for interaction with business systems, which will assist manufacturing organization to better establish, handle, and drive operational data accessibility to the whole enterprise as well as enable to best anticipate the product key performance indicators. Genuinely, the future of PLM does not lie in the growth process. Industry 4.0 revolution is creating a compelling requirement for PLM systems to go beyond the conventional 3D CAD documents monitoring duty and begins to manage the full product consisting of embedded software application and electronic devices through services board on the digital engineering journey, make organization to stay competitive by connecting the real-time product and manufacturing efficiency with their smart connected products.

TECHNOLOGIES OF INDUSTRY 4.0

The journey of PLM in current manufacturing situation is caught up in exactly how to achieve the vision of digitalization. Manufacturing is transforming with the development of Industry 4.0 and it, opens up a new standard of brand-new electronic industrial modern technology, helps in change of the business referred to as a new stage in the industrial change that focuses heavily on interconnectivity, automation, artificial intelligence, and real-time data. Modern technologies such as artificial intelligence, IoT, augmented reality (AR)/virtual reality, additive manufacturing (AM) – also called as 3D printing, simulation, IIoT, and robotics have commonly been talked about as drivers of rapid progress in the manufacturing sector. It encourages manufacturers to far better control and understand every facet of operations and permits them to utilize instant information to improve productivity, boost process, and drive growth. Product managers need to collaborate with plant managers in expanding PLM capabilities with operations to ensure increased coverage of the product lifecycle and stronger integration of processes along the entire value chain.

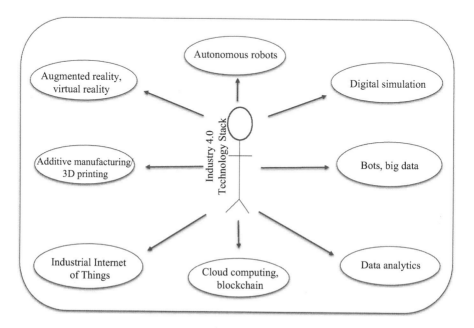

FIGURE 6.1   Industry 4.0 technology stack.

ADDITIVE MANUFACTURING OR 3D PRINTING

AM or 3D printing is a trending subject in engineering, manufacturing, and product development. AM is a technological innovation method by which products are purchased, designed, created, and distributed. AM is used to create models and prototypes from non-production to all set materials. AM is an appropriate name to explain the modern-day innovations that build 3D objects by including layer-upon-layer of material. Sporadically called quick prototyping, 3D printing is an approach of manufacture where layers of a material are established to develop a solid thing. It provides a brand-new surprising feature of the digital thread between engineering and manufacturing that provides the adaptability to alter processes across manufacturing locations to minimize cost and make best use of earnings for the organization.

Representative to AM technology is using a computer system, 3D modeling CAD system, and AM device tools in addition to layering material. When a CAD model is created, the AM device reviews the information in the CAD files and lays down or adds successive layers of fluid, powder, sheet material, or other in a layer-upon-layer style to fabricate a 3D component. The AM application is infinite; the utilization of AM as a kind of quick prototyping is focused on preproduction visualization models. AM is being used in the making of end-use products in aircraft, dental repair services, clinical implants, vehicles, and designing new products as part of new product development.

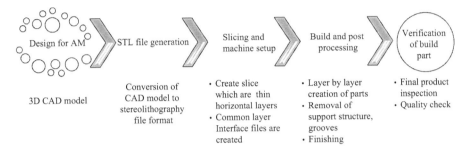

FIGURE 6.2   Stages of additive manufacturing (AM).

The designing of a CAD model is an initial step in the AM process. Reverse engineering can additionally be made use of to create a digital model using 3D scanning. An important phase in the AM process that varies from regular manufacturing technique is the demand to transform a CAD model into an STL (stereolithography) information. STL utilizes triangles (polygons) to specify the surfaces of a thing. A 3D printing machine normally consists of numerous small complex parts; so suitable maintenance with calibration is important to creating specific printing. The removal of the print is as basic as splitting the released part from the construct system. A great deal of 3D printing products have the ability to be sanded, and also numerous other postprocessing techniques, including rolling, high-pressure air cleaning, brightening, and tinting are applied to prepare a print for end use. By combining deposition with machining in a CNC machine, complicated forms can be created and finished with maximum precision; however, by utilizing AM, the assembly can be produced as one single part in a solitary action, while being made lighter and more resilient at the very same time. The complementary customization that AM uses enables for fast prototyping, although fundamental design can be changed to customer needs.

Additive Manufacturing Greets PLM

PLM provides the genuine value chain of integration of BOM with SCM through ERP platforms that provides roadway to crossbreed digital commercial enterprises. When making use of conventional detractive manufacturing strategies, a design layout that would certainly have been produced by reasonably simple setups may need a lot of independently generated components assembled, which will certainly be adding cost and have an impact on restricting design choices. The new format of data is extracted from CAD model, which is required when developing complex shape parts in AM. Input for the AM can be obtained from the product structure maintained in PLM with its own phases of 3D printed component throughout its lifecycle, which plays an important role in the NPD and manufacturing processes. This unique design consideration creates added versions of CAD design. PLM systems are called for to take care of both variations individually to enhance traceability.

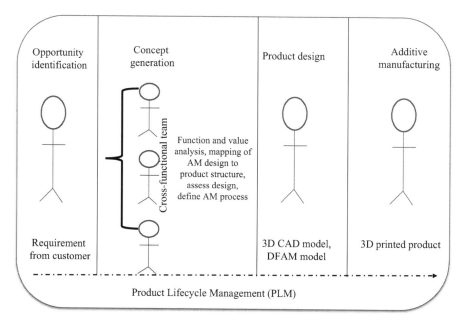

FIGURE 6.3   AM in the phase of PLM.

The developments in CAX combined with PDM and PLM software applications are driving the need for AM. More enhanced, complicated shapes and the need for an extra automated production practice make AM optimal for an expanding variety of manufacturing demands. PLM is everything about innovation and also with the help of AM, it can give better products to consumers. By managing 3D printing data effectively, manufacturers can generate quality products in less time and low cost. AM will dramatically transform the way the product development is done; with a very considerable impact on PLM, as AM is supporting in design chain and supply chain, enterprises are moving toward fast and less expensive way of part prototype through AM innovations. In NPD, AM presents a completely new body of knowledge to DTC together with DFM, disrupting the balance of power between product owners and process owners in collaboration with suppliers, vendors, and customers; thus, PLM provides a joint capability to improve the co-innovation process and speed up the use of AM as a mainstream manufacturing technique for commercial components.

BoT

The BoT is a brand-new revolutionary change with manufacturers in the realms of Industry 4.0. With the innovation in technology, manufacturing organizations have a tendency to select robot growth services as part of the venture remedy. A BoT is a computerized tool that makes the task simpler. The robots have the ability to comprehend the natural languages and are capable of recognizing and also implementing key phrases based on them. It is an automatic program running over the enterprise applications, whereas robotics are mechanical. Yet both excel at repetitive tasks in day-to-day activities. BoTs have two extra attributes that distinguish them from standard enterprise software program. They are goal-oriented and are reactive to the environment. They are enriched by artificial intelligence, natural language handling, and the capability to process disorganized data, so they can act as cognitive virtual agents to function in the direction of desired outcome.

BoTs and robotics both do the same thing. BoTs are software programs and robotics are mechanical. Robot process automation (RPA) or BoTs are significantly efficient in executing tasks that earlier might have been done by humans. RPA is an online software program "robotic" or application that recreates the activities of a human being engaged with the interface as a human would. RPA encompasses not just repeated jobs but also several other sophisticated functions. RPA executes a substantial share of repetitive back office processes in much less time and with more reliability and better compliance. When automation becomes "smarter" and a lot more common, it is paramount that the humans interact with the controlled systems in a secure environment.

RPA Elevates PLM

The current NPD process is driven by information and interactions. RPA is being increasingly used in the manufacturing industry as smart PLM process automation to boost effectiveness of PLM. RPA specifically aids in improving process automation by totally reproducing the actions in generating BOM. RPA supports MES/legacy application and in combination with PLM it automates information along with electronic operations between the applications. Product development has a good deal of records that are created during the growth; examples of the document are the cook book or the service manual that can be automated utilizing RPA. Automation of audit trail of compliance information for each product within the PLM system is accomplished with RPA. RPA is an excellent selection to PLM data migration.

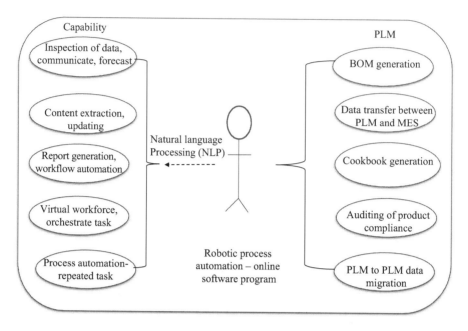

FIGURE 6.4   RPA capability and its role in PLM.

Cobots

Smart manufacturing is driven by dramatic increases in data volumes, computational power, connectivity, machine–human–machine communication, and machine–machine communication, introduction of data analytics and business intelligence, and renovations in bridging the flow between digital twins and the physical world. Classic machine–machine communication forms the foundation; objects generally not considered as machines are attached to a central server or database, and programmable logic controllers synchronized with the actions of robots in the shop floor are automated and taken into consideration as machines within the factory as smart connected ecosystem. Autonomous robots are smart robots that complete jobs with safety and security, adaptability, convenience, and reliability. The smart robotics or collaborative robots (Cobots) operate in direct collaboration with humans inside a defined work area in an organized operation within closed locations. With the introduction of AI, robotics perform jobs as well as process by themselves without the need of human treatment, making them a self-governing entity. Monitoring and data analysis together with the process of simplifying the manufacturing setup process have enhanced its existence in smart equipment with the help of machine learning and 3D printing.

Industry 4.0 advancement has enabled robotics to work with human beings hand in hand in interlinking tasks and also using wise sensor human–machine interfaces in industrial environments. Industrial robots have the ability to dramatically enhance an item's high quality. Applications are executed with accuracy as well as superior repeatability on every task. A lighter, mobile plug-and-play generation of cobots is inward bound on the manufacturing floor to function safely along with human workers, thanks to developments in sensing unit and digital mockup, edge processing, and computing. The cobots are usually used when a certain process involves an additional operation that needs to be executed in the presence of a human. For security functions, these robotics are furnished with security monitored stop system that enables the robotics to stop its operations when it discovers human existence within a predetermined secure area.

The design, simulation, optimization, and offline programming of robotic applications are kept and maintained in application lifecycle management (ALM) to make best use of traceability, adaptability, and also the operation of automated systems with an in-built 3D CAD model; the software program incorporates the simplicity to maximize robotic paths as well as improve cycle times with the power to simulate online mockups of production cells and systems. An exciting objective, in terms of smart robot development, is that the tasks use separate and integrated processes to coordinate product development use PLM for hardware management along with ALM system for software program data and process management. As smart robot manufacturers have become progressively dependent upon software program for including new functions, ALM software has taken over the monitoring of complexity in terms of not just the software program but also the component growth that normally falls under the PLM environment.

Another example of how PLM process will handshake with cobots is the release of manufacturing process instructions combined with shop floor activity. Cobots have accessibility to programming or typical CAD/CAM programs, improving the robotic programs to be maintained in the ALM system. The cobot has the ability of doing high-quality assessment of manufactured components. The process normally includes the full evaluation of finished components, high-resolution images of machined components for precision, and part verification against CAD versions maintained in PDM. The evaluation of components are recorded electronically and digitize the comparison to computer system produced model process and the outcomes are checked out in PLM system resulting in even more exact manufacturing batches.

SIMULATION

The increasing need for smart products in this connected world depending on cyber/physical systems and especially on software application calls for even more prevalent use of simulation, done routinely throughout the product design cycle and not simply at the back end for recognition. PLM within Industry 4.0 is a totally digitized service model with enhanced manufacturing process, where challenging industrial globalization progress toward lowering the cycle time between design conceptualization and performance validation is feasible only with incorporated computer-aided engineering (CAE) or simulation. Industry 4.0 smart product design strategy ensures the flow of simulation information and its results are commonly available to the remainder of the NPD team with the involvement of PLM system. The smart connected digital environment concentrates on using simulation methods to production, systems, and its process.

Engineering organizations are progressively open to collaborate by means of handling simulation process data, not as a separated silo but instead as part of the total development process. The simulation work is treated as an essential element of a business's total PLM approach. Simulation is a standard method for forecasting and assessing the efficiency of smart connected systems that are critically hard to deal. PLM is concentrated on spanning and linking all stages of a product's lifecycle, while CAE simulation is generally restricted to a little but a vital segment within product design. The use of CAE-based analysis helps product designers and manufacturers to decrease risk, and with digital mockup easily available and accessible to the rest of the enterprise, improvements will certainly be understood in many areas of the NPD/NPI.

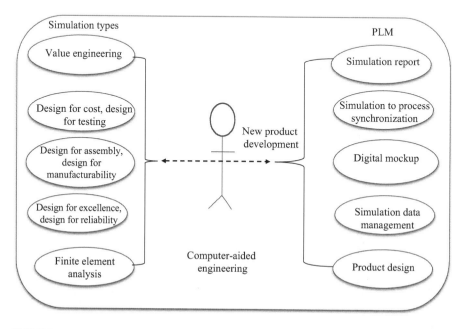

FIGURE 6.5   Simulation data management in PLM.

Simulation procedures and related information are handled in the PLM system in the product development process until the product obtains maturity. Engineering teams get benefit by being able to accelerate simulations and make use of that learning to more effectively drive the product growth. The NPD team lack exposure and also battle to acquire understanding as CAE group provide late results affecting the design information, where simulation compiles a systemic strategy to multi-disciplinary corrective systems with enhanced range of applications. As the customer demands are transforming substantially resulting in product versions, personalized smart connected products require more versatile production systems with increased incorporation of simulation in the product lifecycle management supporting manufacturing operations and service data management.

The digital twin is mainly used in MRO. Digital systems have a tendency to be more autonomous and also are an appropriate decision-maker, assisting people in real time and controlling data in a most efficient way. The electronic depiction of the actual equipment can continuously contrast real determined efficiency versus the twin's substitute performance as well as highlight discrepancy trends that may indicate a requirement for recalibration, or be the precursors of upcoming failure. Factory simulation creates a digital twin for the production systems on a shop floor.

Augmented Reality

Industrial applications of AR, for instance, draw on the digital twin to sustain the maintenance, repair, and overhaul (MRO) of equipment on the factory floor. AR is among the largest modern-day innovation patterns; AR-equipped cellphones in addition to other tools end up being much more obtainable around the world. AR enables us to see the real-life ambience in front of us. In nonspecialist terms, AR is an advanced variant of reality where real-time direct or indirect views of physical real-world situations are shown with computer-generated pictures over a consumer's view of the real world, thereby improving one's current understanding of the fact.

AR-based systems sustain a range of solutions, such as picking elements from a storage facility as well as sending out or dealing with guidelines over mobile devices. These systems are presently in their early stage, although in the future companies will absolutely make much bigger use of AR to provide workers with real-time information to improve services in manufacturing. AR adds electronic aspects to an online sight by making use of the electronic camera on a smart device. Examples of AR experiences include Snapchat and lenses along with the videogame Pokemon Go. AR is rapidly broadening in charm owing to the fact that it brings part of the digital globe into our actual globe, improving the factors the humans see, listen to, and feel. In contrast to numerous other modern innovations, increased fact hinges on the center of the combined fact variety; that is, between the actual world and the digital world. Unlike virtual reality where data need to reside in an entirely digital setup, AR uses existing real environment in addition to superimposing digital information along with it. As both digital and real worlds harmoniously exist side-by-side, users of increased fact experience a brand-new as well as enhanced natural world where electronic info is made use of as a device to offer help in daily tasks.

AR as concept to design, AR as layout for manufacturing, AR as layout for maintenance, but where is AR being utilized in the product design lifecycle? New visualization tools lend itself to better visual communication and also enable a brand-new degree of interaction along the lifecycle. Taking PLM into consideration, it comes close to be characterized by its dispersion as well as collaboration with multiple traditional application within the manufacturing business; the keynote is to remove the relevant understanding and permitting NPD team members to respond on any type of given scenario, thanks to understanding simulation methods. The PLM system utilized both CAD versions and the enhanced fact version needed for AR system to sustain data discussion in addition to interoperability component that make certain information removed from PLM application. By automating the CAD model to AR model data conversion, PLM provides tracking of simulation mechanisms for retrieving and adjusting the data to make conventional design review processes that swiftly accomplish error rectification and enhance product experience.

Smart PLM is developed to have proper combination of customer, process, monitoring, and data to have control throughout the lifecycle process of the product. The PLM stage focuses on defining, specifying, and developing the product layout; examination and simulation are done in design stage followed by production, marketing, and maintenance where preserve and support, elimination or disposal are accomplished by client assistance by ways of MRO. Maintenance is the procedure to accomplish reliability, safety, and total efficiency of the product. Physical components along with digital models are merging even better, producing huge quantities of information and also opening up new vistas for manufacturing as well as maintenance. The AR simulation system runs a real-time virtual representation of the asset in order to understand its performance, so that electronic model may give feedback to the product layout process itself, assisting to optimize design and manufacturing performance in industries such as oil and gas, automotive, aviation, for example, aircraft engine design at an early phase, and so on.

Industry 4.0 simulation technology also raises need for MRO with an increasing competitive manufacturing; the need to drive AR simulation adds extra value, optimizing return on the product asset, and entails both extending their life and also keeping them feasible while appreciating regulatory, safety, and various other regulations. From product design to production, quality assurance, and MRO, what new company opportunities does increased reality in manufacturing fetch? Some of the processes will definitely see timely benefit:

- *Manufacturing process instructions*: By using AR simulation, manufacturing process can be made more interactive and easier. Long-term information can be used in an extremely amazing method that AR depictions normally do, which permit the workers to much better involve and also comprehend the important setting up process. Step-by-step standards provided through an AR headset let the assembly progression to operate faster and also properly respond with the existing information.
- *Defect mapping*: Quality assurance of PCB requires human vision to identify defects in the board; however, with AR simulation, overlays of product will be forecasted along with inefficacy and easily be recognized. AR simulation specialists help to determine one of the most minor issues with higher rate and greater accuracy.

BIG DATA/DATA ANALYTICS

Big Data is one of the most popular topics when discussing Industry 4.0 solutions across today's industrial sectors. It is a procedure of analyzing large and different collections of data to discover market trends, consumer preferences, and other details that work to any businesses. Consuming the information quickly to discover and evaluate the data effectively as well properly will be the greatest challenge and opportunity of Industry 4.0. It refers to high-volume, high-speed, and/or high-variety details, properties that require economical, cutting-edge information on processing that enables improved understanding and real-time decision-making, maximizes production quality, saves energy, and facilitates process automation. Data by itself is of no value, but if the information can be analyzed to expose definition, patterns, or understanding about how a product is used by different consumer sectors, then it has remarkable value to product manufacturers.

Organizations have a traditional custom of capturing transactional details. Besides sensing device, information is collected from machines, industrial equipment, oil pipelines, turbines, energy markets, and so on. Data are accumulated in unbelievably high uniformity. Sensor data supply effective details on the functional efficiency of devices and processes. Other instance is what humans make use of in day-to-day tasks in accessing e-business websites, social networking websites, and social media network. Data evaluation will provide understanding on what promotions could draw in people. It comes to pass by considering not only interests the customers have actually personally specified, but also furthermore recognizing what it is that their circle of associates or pals or products has a passion in. Big Data is crucial not with regard to quantity but in relation to what you perform with the details in addition to just how you utilize it to make evaluation in order to benefit service and solution to the enterprise.

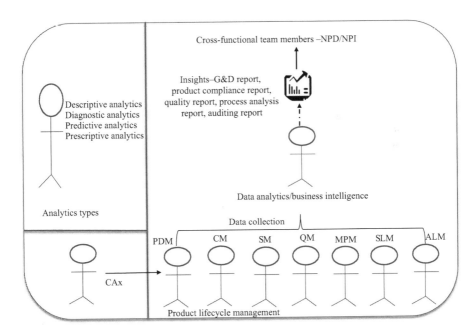

FIGURE 6.6   Data analytics in PLM.

Analytics allow manufacturing organizations to acquire a clear picture of events occurred in the past in addition to the future of their efficiency. Analytics provides the solutions to abide by to "What has actually taken place," "What is taking place" as well as "What will happen." Data analysis tools have really developed with coverage, monitoring, evaluation, and the use of prediction phases. Data analytics is a process that is utilized to take a look at both big and small datasets with differing information to remove purposeful verdicts as well as workable insights. Data analytics has a significant role to play in the development and success of smart product development and smart manufacturing. Analytics tools will enable business devices to make reliable use of the datasets.

One of the best examples of using data analytics is in the field of sports. Industries are fascinated with data capture and wish to have all these data analytics bits and pieces, taking all that information and furthermore drawing out insights out of it. Industries are investing in information specialists efficient in understanding the information along with software and technology to extract it. Data is all over the place. Data analytics in football aids to enhance performance, examine rivals, prevent injuries, and list several of the very best techniques to get additional information in the decision-making process. Using information along with statistics has become prolific throughout most significant manufacturing activities.

Association of Big Data with Data Analytics

With the advancements of new age modern technologies cultivating information and also analytics marvels, the company can process live data and run responses cycle in every action of the product development or operational process. Big Data when partnered with analytics aid company executives to develop origin of falling short in companies, examine process efficiency based on preparing for evaluation for much better decision-making, and sales fads based upon reviewing the consumer buying history. It deals with messed up data resources that have little or no control over its format.

Good data, sound analysis, and beneficial insights are important for reducing threats, making balanced strategic choices, and also resisting versus competitors. Organizations use Big Data to drive vital strategic decisions and also boost business efficiency coupled with analytics. Big Data analytics drive management executives to gear up enterprise decision-makers to make better decisions without endangering the quality of the product and also its service.

Big Data Analytics Clutch Onto PLM

An effective digital journey takes greater than just having access to data; it needs to be presented throughout the whole supply chain and innovative product development process. The PLM system has ample and incredible organization value. Enabling decisions to be made within the PLM system in the early phase and for analytics to be close to the originating system when designing a product is crucial. PLM systems accumulate substantial amounts of information, with most of the product characteristics stored: customer feedback, materials, requirements, and so on. The power of PLM data will be taken advantage of with data analysis, machine learning infuses added intuitions right into the product lifecycle.

Big Data and analytics incorporated with PLM aid NPD/NPI team to develop better products, customized to end-user actions, and new business models. Risks, security, compliance, and regulative needs can be conveniently attached utilizing Big Data analytics. Data analytics will certainly lead manufacturing organization in staying clear of mistakes that have actually run into in the past.

Product layout is enhanced to make use of the least costly production processes possible while still accomplishing product demands of top quality by the clients, by using tolerance evaluation that increases the insurance coverage of what can be examined prior to the shift to production. Merging data analytics with PLM through CAD data management helps to break down the data silos, thereby enabling product managers to specify, share, validate, and evaluate product resistances to ensure that the part specifications are aligned with the manufacturing process. The product designers can examine the impact of tolerances and dimensions on the product design prior to the prototyping of the product through manufacturing by complete integrated 3D tolerance evaluation tool, which gives to the manufacturer invaluable insights into the option of appropriate resistances to accomplish product variant goals successfully. Unlike simple 1D stack-up analysis simulation, information evaluation coupled with product design data is shown in an intuitive graph to be able to produce GD & T report, which is affixed to the design variant and is also available across the product lifecycle.

CLOUD COMPUTING

Cloud computing innovation is thought to be a vital factor in speeding up digital transformation for any organization as an outcome of Industry 4.0. Cloud computing technology makes it possible for organizations that collaborate via the use of a common application or delivery of a solution online hosted on a cloud environment outside the enterprise. Cloud application providers make every effort to offer better service as well as efficiency with the same software application programs as were set up locally within the enterprise. With artificial intelligence and automation being incorporated more frequently into industry, cloud computing is a way for businesses to readily adapt to the technological change. Cloud computing will come as an efficient enterprise tool if utilized properly; organization can save a fair bit of money – as it no longer requires to stress over costly software and hardware purchases; the expenditure can be replaced by subscription model to the cloud applications one requires. In simple terms, end-users have accessibility to cloud-based applications through an Internet browser or a lightweight desktop application or mobile app, whereas the business software application for information are saved on servers at a remote location.

PLM Housed Through Cloud Computing

As an enterprise application, PLM needs a significant financial investment in server and networking infrastructure in addition to an IT team who is educated and available to provide the system and also deliver support. Product development in the smart industrial manufacturing is more probable to be attained by involving networks of geographically distributed vendors/suppliers along with experts, utilizing the necessary ability and competence to make best use of a product's competitive advantage other than prompt delivery. With conventional on-premises enterprise products, PLM sustains substantial licensing over maintenance fees.

The product stakeholder network is a geographically dispersed network of suppliers who all contribute to finishing a final product. As enterprises are moving their focus to a dispersed network, where stakeholders can interact, information is accessible, and there is a shared version of the reality, the cloud PLM system will emerge as the dominant technological service in the supply space. Cloud PLM suggests that individuals can access the same resource of truth at the exact same time from anywhere in the world; it is most likely to be the PLM of the future. Cloud-based PLM is gaining a lot more support owing to its more recent innovation and considerably more progressed IT environment. On-premise PLM implementations are making less sense, both technically and financially. The potential of cloud computing, when correctly rolled out, will have a substantial influence on PLM. The urgent service discomforts and short opportunity will provide an advantage for cloud service providers to use their services to small enterprises as well as OEM at the time they are needed and also in the right amount and with ideal financial savings by eliminating IT expenditures, streamlining execution, scalability, raising protection, and giving a much faster path to ROI.

SUMMARY

Industrial revolution 4.0 urge to combine the digital and physical world in the manufacturing chain that significantly changes the means by which products are being produced. Fast breakthroughs in connection, mobility, analytics, scalability, and information have actually brought about huge improvement. Industry 4.0 will certainly entail retrofitting existing industrialized systems with cutting-edge innovations that will give much more lasting services. The main focus of smart PLM to smart manufacturing is to develop intelligent systems, such as CAX, PDM, M2M, things to things, and HMI along with sensors and communication devices, taking care of the data circulation from smart and furthermore dispersed system collaboration. Integration of advancement of Industry 4.0 technology with PLM offers lots of advantages to manufacturers, including greater performance in product cum process design and development, enhanced rate of manufacturing, and improved product quality.

Manufacturing enterprises that decide on initially will certainly gain a difficult-to-challenge advantage in the race to Industry 4.0. Besides, they will have the ability to establish and influence technological standards for specific businesses. The benefit will by no means be limited to better performances. The genuine objective will certainly be the new business models and revenue streams with which the digital transformation will certainly open a digitized collaborative environment. The implementation of digitized autonomous environment is achieved by discovering individuals with the best skills, together with handling the shift to a culture that is willing to carry out the effort toward transformation.

With Industry 4.0 technologies and Lean Six Sigma technique, organizations will better take advantage of huge amounts of information to make process a lot more efficient to offer better products and services to customers. Collaboration with data insights is the main criterion for the manufacturer to make better choices. The demand for quality assurance and process enhancement is the top most priority. Manufacturing enterprises need to take an evolutionary approach to moving where they are to where they wish to be, and also developing the people together with the innovation will certainly succeed. The real-time QFD is frequently being updated. What added solutions can the manufacturer provide to the consumer? How can they make the product more risk-free and reliable? Smart manufacturing modern technologies have the potential to respond to these inquiries by manufacturing the product more smarter, safely sharing information, and correspondingly connecting the different entities throughout the value chain. Above all, the digital journey gets fulfilled by having a strong business use case and the cost-benefit value, including tangible and intangible returns to enterprises.

BIBLIOGRAPHY

1. Gartner Identifies the Top 10 Internet of Things Technologies for 2017 and 2018. April 4, 2016.
2. "ANSYS Minerva|Simulation Data Management". 2019. Ansys. Com. https://www.ansys.com/products/platform/ansys-minerva.
3. IERC—Standardization Challenges. June 15, 2015. IERC_cluster_book_2012_Web on Internet of Things
4. *Internet Trends 2015—Code Conference*. Mary Meeker. May 27, 2015.
5. M2M standards overview, January 16, 2014 release in November 2013 by Huwaei, www.huwaei.com
6. Tolerance Analysis|Product Brief|Sigmetrix, https://www.sigmetrix.com/products/cetol-tolerance-analysis-software/

7 PLM Using IIoT Use Case

Given the fast-expanding need for smart connected products almost everywhere, suppliers are faced with the complexity of integrating software applications, hardware, firmware, and physical elements and require an understanding of how the enterprise should design, develop, test, and produce products. The customer demands are based on significant experiences across all aspects of this spectacular digital journey. A combination of product lifecycle management along with traditional manufacturing systems such as enterprise resource planning, manufacturing execution systems, and other legacy applications and the Industrial Internet of Things (IIoT) brings a major change in the way the enterprise works. Placing software application designers on the channel of development with their mechanical and electric equivalents delivers extraordinary performance gains. Industries are currently transitioning from product oriented to customer oriented and the techniques are different when it concerns obtaining brand-new and returning consumers. Constructing a client-driven enterprise in today's digital globe is increasingly complex. While new modern technologies have enabled businesses to make quick modifications, consumers now expect greater attention. In simple words, PLM aids in connecting product design with solutions to production along with procedures forming an end-to-end smart connection in the product design, growth, operation, and service.

PLM USE CASE

Products are becoming more advanced and smarter. Enterprises need a better model to support their product development in this competitive era. The PLM manages all aspects of the product lifecycle from concept design to product retirement.

PLM helps enterprises to take a cohesive, holistic view of their products and product-related processes that results in

- efficiency improvements,
- development of new products,
- reduced costs,
- increase in productivity, and
- improved quality of products.

BUSINESS CHALLENGE

Product design is confined only to the design engineers who have access to CAD application. Product design data sharing and nonavailability of the updated version of the CAD model among the design engineers who are working on the same product within the enterprise have been observed across the globe. The approval and review process of design changes are paper based and are accomplished via e-mail or manually.

The history of product-related data tracking is not available and this causes difficulty for product development team. Collaboration of the NPD/NPI team was also lagging.

PRECONDITION

The CAD application for modeling product design should be as per the customer requirement along with NPD/NPI program.

APPROACH

The core of PLM, that is, PDM, helps CAD designers to share design data across counterpart designers with (check-in, check-out) mechanism in common space (central storage) and workspace (personal storage). Product details and information page offer much more understanding to the product structure along with the details concerning "where used by and referenced by." Access control gives stakeholders permission to access product-relevant details based on the role mapped in PLM. The BOM and viewable are different representations of the product structure and its details to get even more clear picture of the product for the NPD/NPI members who does not have access to CAD application. A product would certainly undergo several stages to reach maturity via lifecycle process function. Each of the lifecycle stages have actually different groups included, different service guidelines, and so on. The product go upward (promoted) or backward (demoted) in its maturity. A review or approval is usually made at the end of each phase. Workflows are used to convert business techniques of the firm and/or governing agencies into repeatable, recorded procedures.

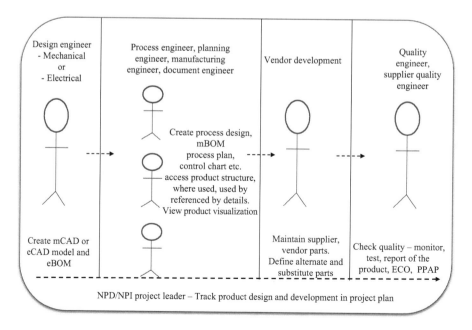

FIGURE 7.1 PLM use case.

RESULT

PLM working along the concurrent engineering principles provides complete configuration control of product data in all phases of development, from initial idea, through design, development, and manufacturing, to complicated product structure, products information, visualization, connected files, engineering changes, and history as well as integration of heterogeneous CAD application with PLM.

PLM–ERP INTEGRATION USE CASE

PLM has actually become vital as the solitary source of reality for handling all facets of the product from initial design to development and manufacturing. The ERP acts as an area to organize reporting relevant to accounting, sales forecasting, client assistance, production planning, and logistics. The objective of the PLM to ERP integration is to transfer all product changes correctly. ERP makes it possible for services to handle besides automate shop floor day-to-day activities. The ERP is used to manage product resources along with financials. It helps with the flow of information across the various divisions, making use of a single resource of information. An ERP focuses on a product lifecycle, when a product has actually been authorized and released. It ensures the supply chain is enhanced; as well as products are made in a timely fashion while making certain accurate price control.

BUSINESS CHALLENGE

An enterprise that implements an ERP system without a PLM system runs the risk of mishandling production changes and also for that reason conducting incorrect monetary preparation as there will certainly be a void in the documents that can make monitoring of the processes extra tough and expensive. Change in one system will certainly influence the other system. Without precise PLM data, the ERP system becomes inadequate.

PRECONDITION

The PLM system and the ERP system should be up and running. It needs to have a business process that will add value to the enterprise on integration of both the systems.

APPROACH

In product design development process, involvement of sources plays a vital role in choosing supplier parts. Components become a part of BOM: it can have parts made in-house or bought out based on the requirement. These data are needed by production planning control members for planning, scheduling, and tracking the shop floor activity. Choosing the right attribute that will have impact on the smooth transformation of business process is the key to successful integration.

Process flow is as follows:

- In-house made parts need to go through vendor development to PPC and get updated in ERP.
- Bought out parts also need to go through vendor development to PPC and get updated in ERP.
- ERP attributes that have been updated in PLM are as MRP and MRP-II attributes such as sourcing type, revision of parts/assembly, customer ID, reference designator, lot number, min order, max order, min–max quantity, UOM, and effective date.

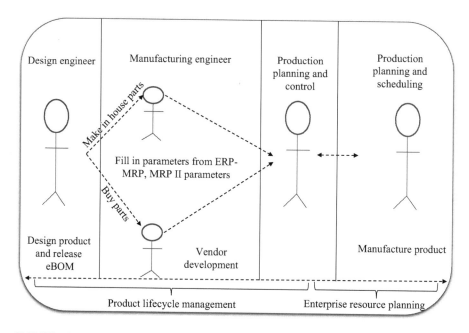

FIGURE 7.2   PLM and ERP integration use case.

RESULT

- Configure/build to customer order
- Increase plant flexibility
- Remove nonvalue added activities
- Streamline customer/supplier collaboration
- Control product cost and quality

Incorporating PLM and ERP systems ensures that both of them have the same data and the most up-to-date details whatsoever times. Information streams are unified and wasted initiatives are eliminated along with maintaining every person on the NPD/NPI team who evaluated the existing condition of a project throughout its development. Strong PLM assimilation will bring about far better enterprise business process optimization that will enhance product development and decision-making processes. The most significant value that derives from the integrity is elimination of multiple entries of same information related to product design and development, which can be made to happen at one system and available across the enterprise.

PLM–ERP–MES INTEGRATION USE CASE

MES is the lifeline of the operations of any manufacturing industry. MES system acts as a digital platform that takes care of the flow of production for final products and provides real-time visibility into shop floor process for improved decision-making. PLM helps in tracking the product design and development process; ERP helps in production planning and scheduling of the product to be manufactured; MES helps in controlling the production planning and scheduling of the production operation in real time. The PLM–ERP–MES combination leverage manufacturers by authorizing it to share design information; planning also allows monitoring of all elements of their manufacturing operation throughout the extended enterprise.

BUSINESS CHALLENGE

PLM as well as ERP is not able to track, control, and keep an eye on the actual makeover of the needed raw materials into the preferred completed products. Some of the main concerns relate to availability of the updated BOM and manufacturing process instructions.

PRECONDITION

PLM and ERP systems should be available at enterprise level, and the MES system should be present at the operation level. All three systems should be up and running with the option enabled for tight integration either unidirectionally or bidirectionally based on the business requirement. It needs to have a strong business use case to automate the process, which will add value to the enterprise on integration of all the systems from design through manufacturing to operations.

APPROACH

A crucial aspect in attaining true performance is NPD/NPI member's self-confidence in the quality of data streaming through the systems. Uncertainty may/will create consistent questioning on fixing data, squandering user, throwing away cash, and time's perseverance. Coordination between NPD and NPI tasks from PLM is required to ensure optimum use of ERP from product launch through production and delivering manufacturing procedure instructions along with mBOM to MES for operation management.

- eBOM is created and released from PLM.
- Manage MRP and MRP-II attributes.
- Create mBOM, manufacturing process instructions, and ECO.
- Through ECO, release the mBOM along with work instructions to ERP.
- Create the required production planning and scheduling in ERP.
- Provide the MES attributes such as work center, production line, plant details, and so on.
- Detailed manufacturing operations are executed in MES.

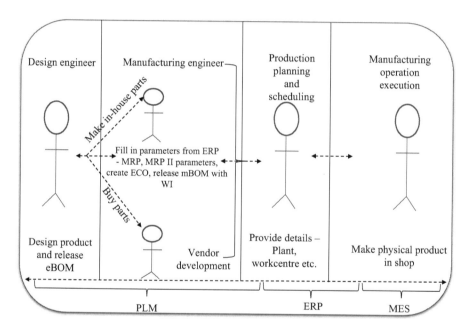

FIGURE 7.3   PLM, ERP, and MES integration use case.

RESULT

By integrating triplets of manufacturing, manufacturers will certainly be aggressive regarding ensuring the client satisfaction by distribution of quality products in a prompt, economical fashion. Work orders together with job direction are pressed immediately to the production line operations besides tracking of top quality of manufactured end products to all NPD/NPI stakeholders in the system.

PLM–IIoT INTEGRATION USE CASE

Collaboration is a major key to manufacturers making it through, besides growing, in this smart connected digital age. Manufacturers are tested by interoperability
along with reverse and correspondingly onward compatibility. The IIoT or Industry
4.0 is changing manufacturing markets with advanced automated digital ecosystem
such as PLM, ERP, and MES, which aids manufacturers to get over the challenges.
Manufacturing working in the smart connected ecosystem will be able to make
higher quality products more effectively and respond faster to changing consumer
demands, thereby fulfilling commitments to consumer and gaining profit.

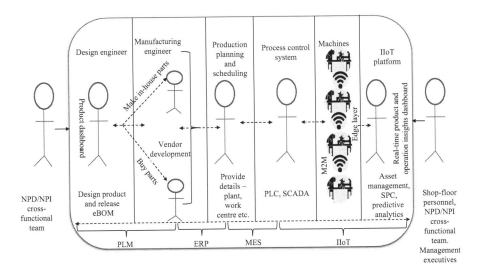

FIGURE 7.4 PLM, ERP, and MES integrated with IIoT.

Business Challenge

It drives tooling efficiency through proper maintenance of the machine tool.

Precondition

- *CAD*: Designing of manufacturing tool
- *CAM*: Design, simulation, and generation of tool path
- *CAE*: Analysis of tool
- *Additive manufacturing*: Tooling prototype
- *PDM and PLM*: Tracking of tool lifecycle, maturity, and other details
- *ERP and MES system*: For process planning and real-time execution
- *PLC*: Control of machine
- *SCADA*: Control and monitor machine tools

Tool wear tracking is done by gauging inner drive signals such as

- Feed electric motor
  - Servo motor
  - Spindle motor
  - Spindle power
- Sensing units as
  - Accelerometer
  - Acoustic emission
  - Thermistor
  - Piezoelectric
- *IIoT platform*: It is used to remotely monitor the functioning of machine tool as well as provide insights into the performance of the tool with predictive analysis.

APPROACH

In manufacturing industries for building a product, the machine tool plays a significant role. Tools have progressed, commonly driven by technological availability, market characteristics, and basic need since the evolution of humans in Mesolithic period. As part of the NPD/NPI, when the product designer starts designing the CAD model, the tool designer simultaneously starts working on the design of tools along with jigs and fixtures based on the requirement. CAE assists in device engineering analysis, optimization of tool design parameters, and motion analysis to analyze the impact of speed, force, stress, and so on. Development in the CAM system helps in capturing the speed, tool path, tool efficiency on the product, and the tool used to carve the product. The generated CAD, CAM, CAE files of the manufacturing tool are stored in the PDM, which will have its own automated lifecycle workflow in PLM taking care of its version, revision, and history of the manufacturing tool lifecycle. The tool design details are passed on to ERP for process planning and further information is passed on to MES for execution of the manufacturing process. Certain number of tools are made readily available in the tool room, which are maintained to take care of continuous manufacturing. Once the tools are fixed in the machine for operations, monitoring of tool wear needs to be taken care of in order to elongate the lifecycle of the manufacturing tool with the involvement of IIoT. Acoustic emission, vibration, temperature, and pressure need to be particularly monitored. Measurements are durable and have been utilized more frequently than any other kind of sensor dimension approaches, and they are a lot more conducive to the commercial field environment. Signals associated with the manufacturing tool wear are captured to keep track of tool health condition. Predictive maintenance helps keep machines running extra effectively and also avoids downtime.

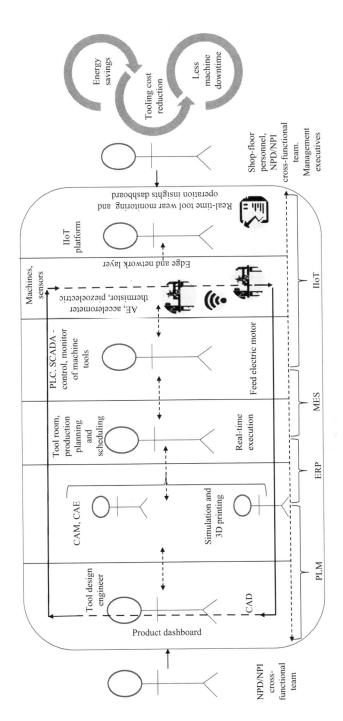

FIGURE 7.5 PLM using IIoT use case.

RESULT

- Energy savings
- Tooling cost reduction
- Less machine downtime
- Increase in operation equipment efficiency
- Increase in productivity of tool room personnel, tool designer, and product designer
- Happy and satisfied customer
- Complete tracking of the manufacturing tool from design through manufacturing to service.

IIoT INVOLVEMENT IN OTHER INDUSTRIES, SENSORS, AND COMMUNICATION

IIoT is involved in many sectors and is not limited to only manufacturing industry.

- *Machine data*: Setup, configuration, status, process
- Machine performance
- *Operational data*: Machine utilization, line utilization, OEE, takt time, and so on.
- Inspection and quality data
- *Traceability*: Consumables, feeder details, and so on
- Material consumption
- Packaging
- *Labs*: DFMA, FEA, testing, and so on
- Inventory
- Facilities
- Safety
- Services

RETURN ON INVESTMENT

Initiating a new business transformation requires a cost-benefit analysis to check whether the initiative is feasibly sound in finance, technology, and resource. The conventional enterprise technology drivers impacting product manufacturers are quality, price, and time. The entire value chain is in full communication with authorized shareholders in real time with full process transparency. Process development is necessary for continual development. The economic advantages will/ought to include the entire organization. Return on financial investment is a popular metric, it can be utilized as a scale of a financial investment's earnings. Industrial sectors are embarking on a new effort of Industry 4.0 modern technology, IIoT, along with other systems that aid product lifecycle. Management executives and key stakeholders of the manufacturing enterprise always have one burning inquiry: "Just how soon will we start seeing some ROI?" One of the keys to measuring gains for a certain objective is to consider both direct and indirect financial savings.

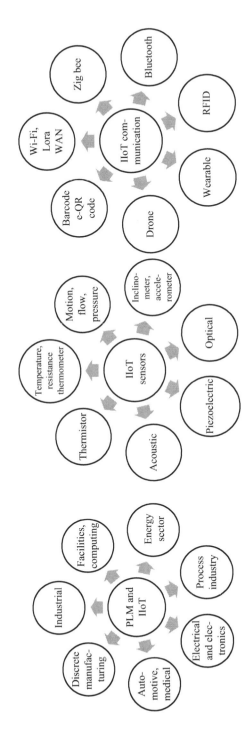

FIGURE 7.6 PLM and IIoT in different industry, IIoT sensors, and IIoT communication.

Manufacturers who are willing to boost the production process and take their enterprise to the subsequent point require to accumulate and assess the pertinent data or appropriate metrics. To overcome the challenge, manufacturer may be secured by performance measurement metrics—Key Performance Indicator (KPI). A KPI is a measurable worth that demonstrates just how successfully a firm is attaining crucial business objectives. Depending upon product development systems in addition to process in the production shop floor, production supervisors may either not be aware of the KPIs that should be tracked to enable them to boost the factory efficiency or they may be incapable to collect sufficient information to properly measure the KPIs that manufacturers want to track. The right metrics will assist the manufacturers to discover the sticking points or weak spots in production floor by offering all the required information and understandings they would need to continually enhance. It is quite crucial to understand the affiliation between high-level goals, purposes, and activities or approaches that are required to achieve them.

The ROI computation needs to take into consideration a huge selection of factors to create a cost design. It must include the time to provision and maintain the applications—PLM, ERP, MES, IIoT platform, servers, network infrastructure, IoT devices, sensors, gateways, and network connectivity devices along with licensing cost, implementation cost, and educating the end users. A ROI can be realized only over the course of years. Simple rule is to define business objectives, proceed with value stream analysis, develop DMAIC (define, measure, analyze, improve and control) model, establish main concern and understanding costs, and have management executives who can drive measurable ROI within their business and monetize investment faster. The typical ROI calculation supplies a measure of a financial investment chance worth in budgetary terms, but other benefits may be equally important to consider. Acknowledging the intangible paybacks of a certain financial investment may help enterprise create an extra detailed growth strategy. Tracking KPIs has actually currently gone much beyond simply picturing details to aid make insightful decisions. Product lifecycle process service applications along with machines in the shop floor are now much more automatic—lowering the amount of decisions required to be made by the stakeholders. KPIs assist the CFT members to keep an eye on the performance of the product and shop floor operation along with machine performance in actual time with very little step in as something exceptional cultivates. I attempted weaving few actions that will aid to compute and comprehend ROI.

Baseline

The suppliers tend to project returns years into the future; the net present value (NPV) discovers the anticipated return these days of financial investment based upon forecasts of future cash money inflows. If the initial price of the investment goes beyond the amount expected to gain from it after inflation over the subsequent years, the financial investment has a negative NPV and will not create value. Begin developing a baseline with the main questions to be answered:

- Establishing realistic expectations based on timing, budgets and resources are very important to ensure that renovation can be pragmatically measured. Besides, goals are iterative and should also be evaluated as well as potentially readjusted on an established timetable to continually drive improvement.
- What are the success indicators for the initiative entailed? Is it product-centric; generate leads; customer-centric; generate income or all?

Baseline represents the minimum rate of return that management executives agree to accept for a connected collective technological investment. Maintaining this base price in mind as enterprise carries out ROI calculations permits stakeholders to eliminate those opportunities that do not fit enterprise financial investment standards.

MEASURE

On setting up the baseline, go ahead with knowing what to measure. It is crucial to maintain enterprise purposes at the leading edge of this procedure, so it helps to gauge what matters and not, what is easy to determine. This may indicate acquiring brand-new analytics tools or establishing new processes to be able to gather information. Consider whether ROI is going to be profit-oriented or it is likely to generate revenue from PLM by means of SLM, smart PLM through smart manufacturing through smart devices. An ROI is a projection of an outcome to take place in the future; a KPI scorecard of metrics is developed to maintain ROI on track.

COLLECT DATA

Use real firm information, not standards, as the return of financial investment is a forecast of what will certainly occur by the end of the time frame. The KPI enables to have real-time insights into the availability of resources. Ideally, there should be a few existences well worth of information consisting of any changes as a result of digital innovation in the product development as well in the operations. Ensure to maintain a clear stock of what business have and what is required.

MANIPULATE DATA

Transforming the data accumulated into a quantitative dimension that can be analyzed prior to making a technology choice allows companies to evaluate the most likely ROI as well as ultimately determine whether a digital collaborative service deserves the investment. As a result of manipulated data, the produced KPI will certainly be able to provide understandings of enterprise efficiency swiftly together with capacity to associate business performance with application efficiency. Prepare a complete roadmap of the improvement, and also the developments are benchmarked at routine periods.

Any digital collaborative transformation initiative has a positive or an unfavorable influence on enterprise. This impact is the "value" that is developed or eliminated. The value can be tangible direct financial advantage or intangible, and it is mostly understood past the closure of the job that drives an initiative. For complete image of value that digital collaborative transformation creates for service, an enterprise requires a measurement that appropriately showcases both tangible and intangible outcomes. This is where return on value (ROV) becomes an essential means of articulating the worth that modern digital innovative technologies campaigns can generate for business. Some of the values are satisfaction level of customer, optimization of product service, increased revenue, increase productivity of stakeholders, and increase in operational equipment efficiency. The ROV shows both abstract and concrete advantages. The foundation for sustained ROI is ROV and it entails a dynamic approach to the driving aspects such as people, customer, and modern technology. No longer would an enterprise see significant ROI without the needed adaptability in adoption, adaptation, and acceptance of a value stream model in business digital transformation.

SUMMARY

Quality is about meeting or going beyond customer expectations. Consumer expectations are increasing and are also competitive. Effectively carried out high-quality processes can minimize rework, scrap, testing, and inspection and boost on-time deliveries. Enterprises must ensure high-quality processes at all stages: from listening to the voice of the customer through reinforcing connected and collaborative NPD/NPI team for producing high-quality smart product with innovative efficiency requirements to ensure that quality criteria report the cross-functional team about the design of products through supply chain to smart manufacturing. Investment in the three main pillars of manufacturing, PLM, ERP, and MES, along with the connected smart manufacturing (IIoT) calls for considerable funding and the benefits will take some time to reap, although smart automation is crucial to stay competitive globally. True value is measured not just in higher profits from greater customer satisfaction but also in higher operational efficiency, functional efficiency, and operation effectiveness.

Start with small business use cases that are specific, measurable, achievable, realistic, and timely, which will provide immediate results in terms of cost, value, and benefit. Develop a proper process planning as it will help in saving cost and time and optimize operation efficiency. Capital funding expenses usually have a tendency to upswing with Industry 4.0 smart connected technology. Proper business strategy planning is a key to develop products and services. Management executives see the advantages in conservation over the long-term possessions, strategic goals where expenses are an ongoing concern together with resources. Preliminary expense is high in connected product design and manufacturing; however, long-term advantages are very effective. Enterprise should consider "smart product to smart manufacturing" as a digital journey that will gradually transform increased functional value across the extended enterprise. The successful transformation of Industry 4.0 within the manufacturing industry will go further ahead with Industry 5.0, and the journey will continue.

Index

Note: Page numbers in italic refer to figures respectively.

9780367431242